Spectral theory and wave operators for the Schrödinger equation

A M Berthier

Université de Paris VI

Spectral theory and wave operators for the Schrödinger equation

Pitman Advanced Publishing Program

BOSTON · LONDON · MELBOURNE

PITMAN BOOKS LIMITED
128 Long Acre, London WC2E 9AN

PITMAN PUBLISHING INC
1020 Plain Street, Marshfield, Massachusetts

Associated Companies
Pitman Publishing Pty Ltd, Melbourne
Pitman Publishing New Zealand Ltd, Wellington
Copp Clark Pitman, Toronto

First published 1982

AMS Subject Classifications: (main) 47B40, 47A40, 35J10
(subsidiary) 42A68, 47A10, 81-XX

Library of Congress Cataloging in Publication Data

Berthier, A. M.
Spectral theory and wave operators for the Schrödinger equation.

(Research notes in mathematics; 71)
Bibliography: p.
Includes indexes.
1. Schrödinger equation. 2. Spectral theory (Mathematics) 3. Differential operators. I. Title. II. Series.
QC174.26.W28B47 1982 515.7′24 82-7653
ISBN 0-273-08562-X AACR2

British Library Cataloguing in Publication Data

Berthier, A. M.
Spectral theory and operators for the Schrödinger equation.—(Research notes in mathematics; 71)
1. Differential operators 2. Spectral theory (Mathematics)
I. Title II. Series
515.7′242 QA329.4

ISBN 0-273-08562-X

ISBN 0 273 08562 X

Reproduced and printed by photolithography
in Great Britain by Biddles Ltd, Guildford

Preface

In the Spring semester, 1978, I taught a course at Rice University on spectral theory. The students were graduate students in Mathematics who had already taken courses on real and complex analysis, integration theory and partial differential equations.

I made detailed lectures notes at that time, and this volume contains a part of those notes with major changes and new material.

The principal aim of this course is to give an almost complete spectral analysis of self-adjoint Schrödinger operators $H = -\Delta + V(x)$ in $L^2(\mathbb{R}^n)$ with short range potentials V. It uses techniques from functional analysis many of which were developed during the last few years. After introducing in Chapters 2, 4 and 5 some basic notions from Hilbert space theory, we discuss the problem of self-adjointness of partial differential operators of Schrödinger type (Chapter 3) and then the existence, the general properties and the asymptotic completeness of the wave operators. The completeness proof given in Chapter 7 is based on stationary arguments. As a preparation for subsequent developments of the theory, we introduce and study in Chapter 8 the restriction of the Fourier transform of certain functions in $L^2(\mathbb{R}^n)$ to spheres in $\mathbb{R}^n$. We shall do this in the language of operators rather than in that of traces, often employed in the mathematics literature. Chapter 9 contains the spectral theory : absolute continuity of the continuous spectrum and bounds on the number of eigenvalues. In particular we give a bound of the Birman-Schwinger type for a non-symmetric factorization of the potential. (Previous treatments of this type of bound depended on a symmetric factorization up to a sign).

The purpose of Chapter 10 is to find a condition on the potential which guarantees that the point spectrum of H contains no point on $(0,\infty)$. The approach we take to this is to reduce it to showing that for a broad class of *periodic* potentials W, the operator $H_p = -\Delta + W$ has no point spectrum.

Indeed we show that $H_p = -\Delta + W$ has purely absolutely continuous spectrum under certain mild restrictions on a periodic potential, following a recent paper by A.M. Berthier. In Chapter 11 we introduce the generalized eigenfunctions associated with the continuous spectrum of Schrödinger operators. We derive continuity properties these functions and obtain the expansion of the vectors in the absolutely continuous subspace of H in terms of the generalized eigenfunctions. Some scattering theory is given in Chapter 13 in order to point out the connection with quantum mechanics (for example the Born approximation, the scattering cross section, and an inverse problem). In particular we include some recent results on scattering cross sections. In Chapter 12 we rederive asymptotic completeness by the so-called "geometrical" method. We also use this method to give an existence proof of the generalized wave operators for Schrödinger operators with long range potentials.

The publication of this book was suggested to me by Salomon Bochner. I wish to express here my deep gratitude for his friendship, his teaching and all the inspiration he gave me.

My thanks also go to the Mathematics Department of Rice University for its kind hospitality. I was fortunate to receive advice from Werner Amrein whose comments helped me to improve this text, and from Peter Perry and to have Mike Taylor read and criticize a large part of it.

I am also indebted to the chairman of my Department at Paris VI, Jean Combes, for his comprehension and support, and to my friend Jocelyne Arpin for excellent typing.

Contents

Le nombre exact demeure racine du symbôle
et dans ton souvenir
ce livre écrit pour toi
Maman "Coubi"
dès l'amère rigueur de ma première
école.

Introduction

The Schrödinger equation is one of the fundamental equations of the physical sciences, as it is the basic equation for the description of quantum phenomena. This is one of the reasons why the so-called Schrödinger operators have received much attention in the mathematical literature.

A Schrödinger operator is an elliptic partial differential operator of the form

$$-\Delta + V(x) \qquad (x \in \mathbb{R}^n) \tag{0.1}$$

or more generally

$$\sum_{j,k=1}^{n} [-i \frac{\partial}{\partial x_j} + W_j(x)]\, a_{jk}(x)\, [-i \frac{\partial}{\partial x_k} + W_k(x)] + V(x) . \tag{0.2}$$

These operators are considered to be acting in the Hilbert space $L^2(\mathbb{R}^n)$, the elements of which have a precise meaning in applications (they are the so-called *wave functions* of quantum mechanics). Under suitable assumptions on the function V(x), one can associate a unique self-adjoint operator H - called the *Hamiltonian* - to the formal expression (0,1). The principal purpose of this course is the mathematical study of such operators H, in particular the determination of the structure and properties of their spectra.

As an example, consider the Hamiltonian H_c describing the motion of an electron in the Coulomb field of a proton. This operator is of the form

$$H_c = -\Delta + \alpha|x|^{-1} ,$$

with $\alpha < 0$, $x \in \mathbb{R}^3$, and acts in $L^2(\mathbb{R}^3)$. H_c has an infinite number of negative eigenvalues that accumulate at the point $\lambda = 0$. These eigenvalues give the possible binding energies of the hydrogen atom ; the eigenfunctions represent the corresponding bound states, i.e. the situation where, roughly speaking, the orbit of the electron is confined to a bounded region in three-dimensional configuration space. Apart from these eigenvalues, the operator H_c has a continuous spectrum extending from $\lambda = 0$ to $\lambda = +\infty$. One may associate a subset of the Hilbert space to this continuous part of the spectrum of H_c ; these wave functions represent the scattering states, in others words unbounded orbits.

In this text we shall prove similar results on the spectrum of $-\Delta + V(x)$ for a whole class of functions $V(x)$ that are either periodic or, like the Coulomb potential $V_c(x) = \alpha|x|^{-1}$, tend to zero as $|x| \to \infty$. In all these cases there occurs a non-empty continuous spectrum. The treatment of the continuous part of the spectrum requires some rather delicate mathematical arguments and will be our principal topic. To motivate the basic idea of the method, we begin in the first chapter by an example that can be explicitely solved. However the operator H that will be considered in this example is not of the Schrödinger type but is simply an ordinary differential operator of the first order.

1 An example

In this first chapter we consider a simple example which illustrates some of the questions that we shall study later. We do not give any proofs in this chapter, and all the notions that we use will be defined precisely later on.

We consider the differential operator $H_o = -i \frac{d}{dx}$ in $L^2(\mathbb{R})$. We denote by $(.,.)$ and $\|.\|$ the scalar product and the norm in $L^2(\mathbb{R})$ respectively, i.e.

$$(f,g) = \int_{-\infty}^{\infty} \overline{f(x)}\, g(x)\, dx \ , \ \|f\| = (f,f)^{\frac{1}{2}} . \tag{1.1}$$

An integration by parts shows that H_o is formally self-adjoint :

$$(f,H_og) = \int_{-\infty}^{\infty} dx\, \overline{f(x)} \ \{-i \frac{dg(x)}{dx}\}$$

$$= -i\, [\overline{f(x)}g(x)]_{-\infty}^{+\infty} + i \int_{-\infty}^{\infty} dx\, \overline{\frac{df(x)}{dx}}\, g(x)$$

$$= \int_{-\infty}^{\infty} dx\, \overline{\{-i \frac{df(x)}{dx}\}} \ g(x) = (H_of,g) \ ,$$

provided that $\lim_{x\to\pm\infty} \overline{f(x)}\, g(x) = 0$. It can be shown that the formal differential operator $-i \frac{d}{dx}$ defines a unique self-adjoint operator in $L^2(\mathbb{R})$. We denote this operator by H_o. This operator has a very simple form after Fourier transformation.

The Fourier transform $\tilde{f}$ of a function f in $L^2(\mathbb{R}) \cap L^1(\mathbb{R})$ is defined as [RS]

$$(Ff)(k) \equiv \tilde{f}(k) = \frac{1}{\sqrt{2\pi}} \int_{-\infty}^{\infty} e^{-ikx} f(x)dx \quad . \tag{1.2}$$

F is a unitary transformation from $L^2(\mathbb{R})$ onto $L^2(\mathbb{R})$. In particular

$$\int_{-\infty}^{\infty} \overline{f(x)}g(x)dx = \int_{-\infty}^{\infty} \overline{\tilde{f}(k)}\tilde{g}(k)dk = (f,g).$$

An integration by parts gives

$$F(-i \frac{df}{dx})(k) = \frac{1}{\sqrt{2\pi}} \int_{-\infty}^{\infty} dx\, e^{-ikx} \{-i \frac{df(x)}{dx}\} = \frac{i}{\sqrt{2\pi}} \int_{-\infty}^{\infty} dx \frac{d(e^{-ikx})}{dx} f(x)$$

$$= k \frac{1}{\sqrt{2\pi}} \int_{-\infty}^{\infty} dx\, e^{-ikx} f(x) = k(Ff)(k) \ ,$$

provided that $\lim_{x \to \pm\infty} f(x) = 0$. Since F is unitary, this shows that H_0 is unitarily equivalent to the multiplication operator by k in $L^2(\mathbb{R})$. One says that the *Fourier transformation diagonalizes* H_0.

We notice some other properties of H_0 :

A) H_0 *has no eigenvalues* :

In fact, the solution of $-i \frac{d}{dx} f(x) = \lambda f(x)$ is $f(x) = c \exp(i\lambda x)$. This function is not square-integrable for any $\lambda \in \mathbb{C}$, so that the operator $-i \frac{d}{dx}$ has no eigenvector in $L^2(\mathbb{R})$.

B) *The spectrum of* H_0 *is continuous and equals* $(-\infty,\infty)$:

The Fourier transformation diagonalizes H_0 as the multiplication operator by the independent variable in $L^2(\mathbb{R})$. This variable assumes all values in $(-\infty,\infty)$, which is essentially our statement.

C) H_o *is an unbounded operator* :

Let χ_n be the characteristic function of the interval $[n,n+1]$, i.e.

$$\chi_n(k) = \begin{cases} 1 & \text{if } n \le k \le n+1 \\ 0 & \text{otherwise .} \end{cases}$$

Then $\|\chi_n\|^2 = \int_{-\infty}^{\infty} |\chi_n(k)|^2 \, dk = \int_n^{n+1} dk = 1$

and $\int_{-\infty}^{\infty} |k \, \chi_n(k)|^2 \, dk = \int_n^{n+1} k^2 \, dk \ge n^2 \int_n^{n+1} dk = n^2$.

Thus, given $M < \infty$, there exists a vector f in $L^2(\mathbb{R})$ with $\|f\|^2 = 1$ and $\|H_o f\|^2 > M$, i.e. H_o cannot be bounded.

We now define for each $t \in \mathbb{R}$

$$U_t = \exp(-iH_o t) \quad . \tag{1.3}$$

This may be written explicitly as follows : If $f \in L^2(\mathbb{R})$, then

$$(FU_t f)(k) = e^{-ikt} \, \tilde{f}(k) \ ,$$

since H_o is just multiplication by k on $\tilde{f}(k)$. Since $e^{-ikt} \, e^{-ik\tau} = e^{-ik(t+\tau)}$, it follows that

$$U_t U_\tau = U_{t+\tau} \quad , \tag{1.4}$$

i.e. the family of operators $\{U_t\}$ forms a *one-parameter group*. In particular $U_t U_{-t} = U_{-t} U_t = U_o = \exp(-iH_o 0) = I$, where I is the identity operator, i.e. each U_t is an *invertible* operator, and $U_t^{-1} = U_{-t}$. In addition each U_t is *unitary* :

$$\|U_t f\|^2 = \int_{-\infty}^{\infty} |e^{-ikt}\, \tilde{f}(k)|^2\, dk = \int_{-\infty}^{\infty} |\tilde{f}(k)|^2\, dk = \|f\|^2 .$$

Thus $\{U_t\}$ forms a *one-parameter unitary group* in $L^2(\mathbb{R})$. By using the inverse Fourier transformation

$$f(x) = (F^{-1}\tilde{f})(x) = \frac{1}{\sqrt{2\pi}} \int_{-\infty}^{\infty} dk\, e^{ikx}\, \tilde{f}(k) ,$$

it is easy to obtain the action of U_t on $f(x)$:

$$(U_t f)(x) = \frac{1}{\sqrt{2\pi}} \int_{-\infty}^{\infty} dk\, e^{ikx} (\widetilde{U_t f})(k)$$

$$= \frac{1}{\sqrt{2\pi}} \int_{-\infty}^{\infty} dk\, e^{ikx}\, e^{-ikt}\, \tilde{f}(k) = f(x-t) \tag{1.5}$$

i.e. U_t just induces a translation by t.

Now consider a second differential operator

$$H = -i \frac{d}{dx} + V(x) ,$$

with $V(x)$ real, bounded and $V \in L^1(\mathbb{R})$. Since V is real, one sees as for H_o that H is formally self-adjoint, i.e. $(f,Hg) = (f,H_o g) + (f,Vg) = (H_o f,g) + (Vf,g)$, because $\int_{-\infty}^{\infty} dx\, \overline{f(x)}\, \{V(x)\, g(x)\} = \int_{-\infty}^{\infty} dx\, \{\overline{V(x) f(x)}\} g(x)$. Define $\rho(x) = \int_0^x V(y)\, dy$ and let W be the multiplication operator by $e^{-i\rho(x)}$, i.e.

$$(Wf)(x) = e^{-i\rho(x)} f(x) . \tag{1.6}$$

Clearly W is a unitary operator, and

$$(W^{-1}f)(x) = e^{i\rho(x)}f(x) .$$

Hence

$$(WH_oW^{-1}f)(x) = e^{-i\rho(x)} [-i \frac{d}{dx} e^{i\rho(x)} f(x)]$$

$$= e^{-i\rho(x)}\{e^{i\rho(x)}.(-i \frac{df(x)}{dx}) + e^{i\rho(x)} \frac{d\rho(x)}{dx} f(x)\}$$

$$= [-i \frac{d}{dx} + V(x)] f(x) = (Hf)(x) ,$$

i.e. $$H = WH_oW^{-1} . \tag{1.7}$$

Thus H is unitarily equivalent to H_o. In particular it has the same spectral properties as H_o (i.e. no eigenvalues and a continuous spectrum equal to $(-\infty,\infty)$). The operator $\mathcal{F}W^{-1}$ diagonalizes H :

$$(\mathcal{F}W^{-1}Hf)(k) = (\mathcal{F}W^{-1}WH_oW^{-1}f)(k)$$

$$= (\mathcal{F}H_oW^{-1}f)(k) = k(\mathcal{F}W^{-1}f)(k) ,$$

i.e. H is multiplication by k on $(\mathcal{F}W^{-1}f)(k)$.

Here we have obtained the spectral properties of H by showing that it is unitarily equivalent to a simple operator H_o the spectral properties of which are known. We have guessed what the unitary operator W is. To treat more general cases, it will not be possible to guess a candidate for W, and therefore one would like to have a *constructive procedure* to obtain such an operator from H and H_o. This is the method of the wave operators which we shall develop in detail later on. One looks at the two unitary one-parameter groups

$$U_t = \exp(-iH_o t) \ , \ V_t = \exp(-iHt) \tag{1.8}$$

and considers the (strong) limits

$$\Omega_\pm = \operatorname*{s-lim}_{t \to \pm\infty} V_{-t} U_t \ , \tag{1.9}$$

if they exist. $\Omega_\pm$ are called the *wave operators* and give a unitary equivalence between H and H_o.

In the above example this can easily be checked. One has

$$\exp(-iHt) = W \exp(-iH_o t) W^{-1} \ ,$$

i.e.

$$[\exp(-iHt)f](x) = e^{-i\rho(x)} [U_t W^{-1} f](x)$$

$$= e^{-i\rho(x)} \, e^{i\rho(x-t)} f(x-t) \ .$$

Since $(U_t f)(x+t) = f(x+t-t) = f(x)$, we have

$$(V_{-t} U_t f)(x) = e^{-i\rho(x)} \, e^{i\rho(x+t)} (U_t f)(x+t)$$

$$= e^{-i\rho(x)} \, e^{i\rho(x+t)} f(x) = e^{i\int_x^{x+t} V(y)dy} f(x) \ ,$$

so that

$$(\Omega_{\pm} f)(x) = \lim_{t\to\pm\infty} (V_{-t} U_t f)(x)$$

$$= e^{i\int_x^{\pm\infty} V(y)dy} f(x) .$$

Thus $\Omega_{\pm}$ are the multiplication operators by

$$\exp(i \int_x^{\pm\infty} V(y)\, dy) .$$

Since V is real, $|\exp(i \int_x^{\pm\infty} V(y)dy)| = 1$, so that $\Omega_{\pm}$ are unitary. Also

$$(\Omega_{\pm}^{-1} f)(x) = \exp(-i \int_x^{\pm\infty} V(y)\, dy) .$$

Hence, as before

$$(\Omega_{\pm} H_0 \Omega_{\pm}^{-1} f)(x) = e^{i\int_x^{\pm\infty} V(y)dy} \{-i \frac{d}{dx} e^{-i\int_x^{\pm\infty} V(y)dy} f(x)\}$$

$$= -i \frac{d}{dx} f(x) - \left[\frac{d}{dx} \int_x^{\pm\infty} V(y)dy\right] f(x)$$

$$= -i \frac{d}{dx} f(x) + V(x) f(x) = (Hf)(x) .$$

It follows again that $F\Omega_{\pm}^{-1}$ diagonalize H.

We shall proceed similarly for Schrödinger operators, where we shall take $H_0 = -\Delta$, $H = -\Delta + V(x)$. The above example shows that we shall have to study the following problems :

a) Prove the self-adjointness of H_0 and H (if H were not self-adjoint, exp(-iHt) would not be unitary, so that $\Omega_\pm$ would not give a unitary equivalence between H and H_0, and the method could fail to give the spectral properties of H).

b) Define the spectrum of a self-adjoint operator and study its subdivisions (e.g. continuous spectrum, eigenvalues).

c) Establish the spectral properties of $H_0 = -\Delta$.

d) Prove the existence of the wave operators and study their properties.

e) Obtain the spectral properties of H.

2 Notions from Hilbert space theory

In this chapter we collect some definitions and results about Hilbert spaces and linear operators. These results will be needed at various places during the course.

DEFINITION : A *separable Hilbert space* $\mathcal{H}$ is a linear vector space over the field $\mathbb{C}$ of complex numbers having the following properties, [K], [AG] :

(a) There exists a strictly positive *scalar product* in $\mathcal{H}$, i.e. a mapping $\mathcal{H} \times \mathcal{H} \to \mathbb{C}$, denoted (.,.), which is linear in the second and antilinear in the first argument. Thus :

$$\|f\| = (f, f)^{\frac{1}{2}} > 0 \quad \text{unless} \quad f = 0$$

$$(f, g) = \overline{(g, f)} \tag{2.0}$$

$$(f, \alpha g+h) = \alpha(f, g) + (f, h) .$$

(b) $\mathcal{H}$ is *complete* in the above norm (2.0) : If $\{f_n\}$ is a Cauchy sequence (i.e. such that $\|f_n-f_m\| \to 0$ as $n,m \to \infty$), then there exists $f \in \mathcal{H}$ such that $\|f_n-f\| \to 0$ as $n \to \infty$.

(c) $\mathcal{H}$ is *separable* i.e. there exists a denumerable sequence $\{f_n\}$ which is dense in $\mathcal{H}$.

DEFINITION : A subset $\mathcal{D}$ of $\mathcal{H}$ is *dense* in $\mathcal{H}$ if, given any $f \in \mathcal{H}$ and any $\varepsilon > 0$, there exists $g \in \mathcal{D}$ such that $\|g-f\| < \varepsilon$.

Two basic inequalities which follow from these axioms are :

the *Schwarz inequality* :

$$|(f,g)| \leq \|f\| \; \|g\| , \tag{2.1}$$

and the *triangle inequality* for the norm $\|.\|$:

$$\|f + g\| \leq \|f\| + \|g\| . \tag{2.2}$$

There are two notions of convergence in $\mathcal{H}$:

DEFINITION :

(a) A sequence $\{f_n\} \in \mathcal{H}$ *converges strongly* to $f \in \mathcal{H}$ if $\lim_{n\to\infty} \|f_n - f\| = 0$. One writes $\text{s-}\lim_{n\to\infty} f_n = f$.

(b) A sequence $\{f_n\} \in \mathcal{H}$ *converges weakly* to $f \in \mathcal{H}$ if $\lim_{n\to\infty} (f_n, g) = (f,g)$ for each $g \in \mathcal{H}$.

One writes $\text{w-}\lim_{n\to\infty} f_n = f$.

The Cauchy criterion is valid for each type of convergence :

(a) if $\{f_n\}$ is strongly Cauchy, $\|f_n - f_m\| \to 0$ as $n,m \to \infty$, there exists $f \in \mathcal{H}$ such that $f = \text{s-}\lim_{n\to\infty} f_n$;

(b) if $\{f_n\}$ is weakly Cauchy, $(f_n,g) - (f_m,g) \to 0$ as $n,m \to \infty$ for each $g \in \mathcal{H}$, there exists $f \in \mathcal{H}$ such that $f = \text{w-}\lim_{n\to\infty} f_n$.
(Notice that (a) is nothing but the completeness of $\mathcal{H}$).

LEMMA 2.1 : Let $\{f_n\}$, $f \in \mathcal{H}$. Then $\text{s-}\lim_{n\to\infty} f_n = f$ if and only if $\text{w-}\lim_{n\to\infty} f_n = f$ and $\lim_{n\to\infty} \|f_n\| = \|f\|$.

This lemma indicates that strong convergence is a stronger requirement than weak convergence. This can be seen in detail in the example below :

DEFINITION :

(a) f is *orthogonal* to g, written $f \perp g$, if $(f,g) = 0$.

(b) Let $\mathcal{D}$ be a subset of $\mathcal{H}$. The *orthogonal complement* $\mathcal{D}^{\perp}$ of $\mathcal{D}$ is defined as

$$\mathcal{D}^{\perp} = \{f \in \mathcal{H} \mid f \perp g \text{ for all } g \in \mathcal{D}\}.$$

Example 2.1 : Let $\mathcal{H}$ be infinite-dimensional, let $\{e_n\}_{n=1}^{\infty}$ be an orthonormal basis of $\mathcal{H}$. If $n \neq m$, then

$$\|e_n - e_m\|^2 = \|e_n\|^2 + \|e_m\|^2 - (e_n , e_m) - (e_m , e_n) = 2,$$

and $\{e_n\}$ is not strongly Cauchy.

Let $g \in \mathcal{H}$. Then g can be expanded in the orthonormal basis $\{e_n\}$ [i.e. $(e_n , e_m) = \delta_{nm}$] and we have

$$g = \sum_{n=1}^{\infty} (e_n , g)e_n \tag{2.3}$$

with
$$\|g\|^2 = \sum_{n=1}^{\infty} |(e_n , g)|^2 \quad , \tag{2.4}$$

where the infinite sum in (2.3) is defined as a strong limit, $\left(\sum_{n=1}^{\infty} \alpha_n e_n = \text{s-lim}_{N\to\infty} \sum_{n=1}^{N} \alpha_n e_n\right)$.

Since $\|g\| < \infty$, it follows from (2.4) that

$$\lim_{n\to\infty} (e_n , g) = 0 = (0 , g)$$

i.e. $\{e_n\}$ converges weakly to zero. Clearly $\{e_n\}$ does not verify the last requirement in Lemma 2.1, i.e.

$$\lim_{n \to \infty} \|e_n\| = \lim_{n \to \infty} 1 = 1 \neq \|0\| = 0 \quad .$$

DEFINITION :

(a) A *subspace* M of $\mathcal{H}$ is a subset of $\mathcal{H}$ which is itself a Hilbert space.

(b) A *linear manifold* is a linear subset of $\mathcal{H}$.

Example : Let N be a linear manifold. Then its closure $\overline{N}$ (in the norm $\|.\|$) is a subspace of $\mathcal{H}$.

Projection Theorem : Let M be a subspace of $\mathcal{H}$. Then

$$\mathcal{H} = M \oplus M^{\perp} \quad ,$$

i.e. each $f \in \mathcal{H}$ may be written as $f = f_1 + f_2$ with $f_1 \in M$, $f_2 \in M^{\perp}$, [K].

DEFINITION : A *linear operator* A in $\mathcal{H}$ is a linear mapping defined on a linear manifold D(A) with values in $\mathcal{H}$. D(A) is called the *domain* of the operator A.

Thus $A : D(A) \to \mathcal{H}$ with $A(\alpha f + g) = \alpha Af + Ag$.

DEFINITION : A linear operator A is *bounded* if there exists $M < \infty$ such that

$$\|Af\| \leq M \|f\| \quad \text{for all} \quad f \in D(A) \; .$$

The *norm* $\|A\|$ of A is then defined as the infimum of all possible values of M : $\|A\| = \inf_{\substack{f \in \mathcal{H} \\ f \neq 0}} \frac{\|Af\|}{\|f\|}$.

<u>DEFINITION</u> : $\mathcal{B}(\mathcal{H})$ denotes the set of all bounded and everywhere defined operators.

It follows from the triangle inequality for vectors that, for $A,B \in \mathcal{B}(\mathcal{H})$:

$$\|A + B\| \leq \|A\| + \|B\| . \tag{2.5}$$

Furthermore, from the definition of $\|A\|$, one obtains for $A,B \in \mathcal{B}(\mathcal{H})$:

$$\|AB\| \leq \|A\| \, \|B\| . \tag{2.6}$$

<u>DEFINITION</u> : Let $\{A_n\}_{n=1}^{\infty}$ be a sequence of operators in $\mathcal{B}(\mathcal{H})$ and let $A \in \mathcal{B}(\mathcal{H})$.

(a) $\{A_n\}$ *converges weakly* to A, written $\underset{n\to\infty}{\text{w-lim}}\, A_n = A$, if $\{A_n f\}$ converges weakly to Af for each $f \in \mathcal{H}$.

(b) $\{A_n\}$ *converges strongly* to A, written $\underset{n\to\infty}{\text{s-lim}}\, A_n = A$, if $\|A_n f - Af\| \to 0$ for each $f \in \mathcal{H}$.

(c) $\{A_n\}$ *converges uniformly* (or *in norm*) to A, written $\underset{n\to\infty}{\text{u-lim}}\, A_n = A$, if $\|A_n - A\| \to 0$ as $n \to \infty$.

<u>LEMMA 2.2</u> :

(a) If $\underset{n\to\infty}{\text{s-lim}}\, A_n = A$ and $\underset{n\to\infty}{\text{s-lim}}\, B_n = B$, then $\underset{n\to\infty}{\text{s-lim}}\, A_n B_n = AB$.

(b) If $\underset{n\to\infty}{\text{u-lim}}\, A_n = A$ and $\underset{n\to\infty}{\text{u-lim}}\, B_n = B$, then $\underset{n\to\infty}{\text{u-lim}}\, A_n B_n = AB$.

[$\text{w-lim}_{n\to\infty} B_n = B$ and $\text{w-lim}_{n\to\infty} A_n = A$ does not imply $\text{w-lim}_{n\to\infty} A_nB_n = AB$, for example].

Proof :

(α) The principle of uniform boundedness implies that $\{A_n\}$ is bounded, i.e. there exists $K < \infty$ such that $\|A_n\| \le K$ for all $n = 1,2,\dots$.

(β) Let $f \in \mathcal{H}$. Then

$$\begin{aligned}\|A_nB_nf - ABf\| &= \|(A_n - A)Bf + A_n(B_n - B)f\| \\ &\le \|(A_n - A)Bf\| + \|A_n\|\,\|(B_n - B)f\| \\ &\le \|(A_n - A)Bf\| + K\|(B_n - B)f\| .\end{aligned}$$

Given $\varepsilon > 0$, we may choose N such that for $n > N$, $\|(A_n - A)Bf\| < \varepsilon/2$, $\|(B_n - B)f\| < \varepsilon/2K$, whence

$$\|A_nB_nf - ABf\| < \varepsilon \quad \text{for all} \quad n > N \quad .$$

Hence $\text{s-lim}_{n\to\infty} A_nB_nf = ABf$ for each $f \in \mathcal{H}$.

(γ) The proof with the uniform limit is similar. ∎

Uniform convergence implies strong convergence : Since for each f in $\mathcal{H}$, $\|Af\| \le \|A\|\,\|f\|$ we have

$$\|Af - A_nf\| \le \|A - A_n\|\,\|f\| \to 0 , \quad \forall f \in \mathcal{H}.$$

Strong convergence implies weak convergence by Lemma 2.1 but the converse is not true.

The following theorem generalizes the geometric series for $(1-x)^{-1}$:

PROPOSITION 2.3 : (Neumann series) - Let $A \in \mathcal{B}(\mathcal{H})$ and $\|A\| < 1$. Then $(I - A)$ is invertible, $(I - A)^{-1}$ belongs to $\mathcal{B}(\mathcal{H})$ and is given by the uniformly convergent series

$$(I - A)^{-1} = \sum_{n=0}^{\infty} A^n = I + A + A^2 + \dots .$$

Also $\|(I - A)^{-1}\| \leq (I - \|A\|)^{-1}$.

Proof :

(i) $\|A^n\| = \|AA^{n-1}\| \leq \|A\| \, \|A^{n-1}\| \leq \dots \leq \|A\|^n$. Since $\|A\| < 1$, this implies $\|A^n\| \to 0$ as $n \to \infty$, i.e. $\underset{n \to \infty}{\text{u-lim}} A^n = 0$.

(ii) Now from the triangle inequality and (i)

$$\|\sum_{k=n}^{m} A^k\| \leq \sum_{k=n}^{m} \|A^k\| \leq \sum_{k=n}^{m} \|A\|^k = \|A\|^n \sum_{k=0}^{m-n} \|A\|^k$$

$$= \|A\|^n \frac{1-\|A\|^{m-n}}{1 - \|A\|} \leq \|A\|^n \frac{1}{1 - \|A\|} \to 0 \text{ as } n \to \infty. \qquad (2.7)$$

This means that $\underset{n \to \infty}{\text{u-lim}} \sum_{k=0}^{n} A^k$ exists and thus defines an operator in $\mathcal{B}(\mathcal{H})$.
By setting $n = 0$ in (2.7) one also obtains

$$\|\sum_{k=0}^{\infty} A^k\| \leq \frac{1}{1 - \|A\|} .$$

(iii) One has

$$I - A^n = (I+A+A^2+\dots+A^{n-1})(I - A) = (I - A)(I+A+A^2+\dots+A^{n-1}) .$$

By taking the uniform limit on both sides, we get

$$I = \underset{n\to\infty}{\text{u-lim}}\,(I - A^n) = \left(\sum_{k=0}^{\infty} A^k\right)(I - A) = (I - A)\left(\sum_{k=0}^{\infty} A^k\right) . \qquad (2.8)$$

The second identity in (2.8) [i.e. $I = (I - A) \sum_{k=0}^{\infty} A^k$] means that $(I - A)$ is surjective. The first identity says that $(I - A)$ is invertible and that $\sum_{k=0}^{\infty} A^k$ coincides with $(I - A)^{-1}$ on the range of $(I - A)$. Since $(I - A)$ is surjective, its range is $\mathcal{H}$, i.e. $(I - A)^{-1}$ is defined everywhere in $\mathcal{H}$ and

$$(I - A)^{-1} = \sum_{k=0}^{\infty} A^k \ . \ \blacksquare$$

DEFINITION : An operator $T \in \mathcal{B}(\mathcal{H})$ is a *finite rank operator* if it can be written as

$$Tf = \sum_{i=1}^{N} (g_i , f)\, h_i \ ,$$

where $\{g_i , h_i\}$ are 2N vectors in $\mathcal{H}$ and $N < \infty$.

Notice that the range of T is the finite-dimensional subspace spanned by $\{h_1,\ldots,h_N\}$.

The set $\mathcal{B}_F$ of all finite-rank operators is a two-sided $*$-ideal in $\mathcal{B}(\mathcal{H})$, i.e. if $T, T_1 \in \mathcal{B}_F$, $B \in \mathcal{B}(\mathcal{H})$ and $\alpha \in \mathbb{C}$, then

(a) $T^* \in \mathcal{B}_F$.

(b) $BT \in \mathcal{B}_F$ and $TB \in \mathcal{B}_F$.

(c) $T + \alpha T_1 \in \mathcal{B}_F$.

DEFINITION : An operator $A \in \mathcal{B}(\mathcal{H})$ is *compact* if there exists a sequence $\{T_N\}$ of finite-rank operators such that $\|A - T_N\| \to 0$ as $N \to \infty$.

Remark 2.2 : An equivalent definition is : A is compact if, for every bounded sequence $\{f_n\}$ in $\mathcal{H}$, the sequence $\{Af_n\}$ has a strongly convergent subsequence.

DEFINITION : $\mathcal{B}_\infty$ denotes the set of all compact operators.

LEMMA 2.4 : $\mathcal{B}_\infty$ is a uniformly closed two-sided $*$-ideal in $\mathcal{B}(\mathcal{H})$:

(a) If $A_1, A \in \mathcal{B}_\infty$, $B \in \mathcal{B}(\mathcal{H})$ and $\alpha \in \mathbb{C}$, then $A^* \in \mathcal{B}_\infty$, $AB \in \mathcal{B}_\infty$, $BA \in \mathcal{B}_\infty$ and $A + \alpha A_1 \in \mathcal{B}_\infty$.

(b) If $\{A_n\} \in \mathcal{B}_\infty$, $A \in \mathcal{B}(\mathcal{H})$ and $\lim_{n\to\infty} \|A - A_n\| = 0$ then $A \in \mathcal{B}_\infty$.

To verify that an operator is compact in an L^2-space, one usually tries to approximate it by a Hilbert-Schmidt operator. The details are as follows :

DEFINITION : An operator $A \in \mathcal{B}(\mathcal{H})$ is a *Hilbert-Schmidt* operator if

$$\|A\|_{HS}^2 \equiv \sum_{k=1}^{\infty} \|Ae_k\|^2 < \infty,$$

where $\{e_k\}$ is an orthonormal basis of $\mathcal{H}$. The set of all Hilbert-Schmidt operators is denoted by $\mathcal{B}_2$.

It can be shown that $\sum_{k=1}^{\infty} \|Ae_k\|^2$ is the same for each orthonormal basis $\{e_k\}$, so that the Hilbert-Schmidt norm $\|A\|_{HS}$ depends only on the operator A, [K].

LEMMA 2.5 :

(a) $\mathcal{B}_2$ is a two-sided $*$-ideal in $\mathcal{B}(\mathcal{H})$.

(b) Every Hilbert-Schmidt operator is compact, i.e.

$$\mathcal{B}_2 \subseteq \mathcal{B}_\infty.$$

(c) $$\|A\|_{HS} = \|A^*\|_{HS} \quad .$$

(d) $$\|A\| \le \|A\|_{HS} \quad .$$

Proof : We prove only (b) and (d) :

(d) Fix $f \neq 0$, $f \in \mathcal{H}$. Choose $\{e_k\}$ such that $e_1 = \dfrac{f}{\|f\|}$. Then

$$\frac{\|Af\|^2}{\|f\|^2} = \|Ae_1\|^2 \le \sum_k \|Ae_k\|^2 = \|A\|_{HS}^2 \quad .$$

Hence

$$\|A\| = \sup_{f \neq 0} \frac{\|Af\|}{\|f\|} \le \|A\|_{HS} \quad .$$

(b) Let $A \in \mathcal{B}_2$ and $\{e_k\}$ be an orthonormal basis of $\mathcal{H}$. For each $N < \infty$, define T_N by

$$T_N f = \sum_{i=1}^{N} (e_i , f) A e_i \quad .$$

Clearly $T_N \in \mathcal{B}_F$, $T_N e_k = 0$ if $k > N$, $T_N e_k = Ae_k$ if $k \le N$. Then by (d), we have

$$\|A - T_N\|^2 \le \|A - T_N\|_{HS}^2 = \sum_{k=1}^{\infty} \|Ae_k - T_N e_k\|^2$$

$$= \sum_{k=N+1}^{\infty} \|Ae_k\|^2 \to 0 \quad \text{as} \quad N \to \infty,$$

since $\sum_{k=1}^{\infty} \|Ae_k\|^2 < \infty$. Thus $A \in \mathcal{B}_\infty$ by the definition of a compact operator. ∎

An integral operator A in $L^2(\Sigma, d\mu)$ is given by

$$(Af)(x) = \int_\Sigma K(x,y)\, f(y)\, d\mu(y) ,$$

where K is a μ-mesurable function from $\Sigma \times \Sigma$ into $\mathbb{C}$.

PROPOSITION 2.6 : Let $\mathcal{H} = L^2(\Sigma, d\mu)$. An operator A in $\mathcal{B}(\mathcal{H})$ is Hilbert-Schmidt if and only if it is an integral operator with square integrable kernel satisfying

$$\int_\Sigma \int_\Sigma d\mu(x)\, d\mu(y)\, |K(x,y)|^2 < \infty.$$

One has

$$\|A\|_{HS}^2 = \int_\Sigma \int_\Sigma d\mu(x)\, d\mu(y)\, |K(x,y)|^2 .$$

We omit the proof and consider the special case $\mathcal{H} = L^2(\mathbb{R}^n)$:

PROPOSITION 2.7 : Let $\phi, \psi : \mathbb{R}^n \to \mathbb{C}$ belong to $L^2(\mathbb{R}^n)$. Let A be the integral operator with kernel $\phi(x)\, \tilde{\psi}(y-x)$. Then $A \in \mathcal{B}_2$ and $\|A\|_{HS} = \|\phi\|_2 \, \|\psi\|_2$.

Notations : $\|f\|_p$ means the L^p-norm of f .

$\tilde{\psi} \equiv F\psi$ denotes the Fourier transform of ψ :

$$\tilde{\psi}(\xi) = (F\psi)(\xi) = (2\pi)^{-n/2} \int dx\, e^{-i\xi . x}\, \psi(x) . \tag{2.9}$$

Proof : This follows from Proposition 2.6. We use the change of variable $y \mapsto z = y - x$:

$$\|A\|_{HS}^2 = \int dx\, dy\, |\phi(x)|^2\, |\tilde{\psi}(y-x)|^2$$

$$= \int dx\, dz\, |\phi(x)|^2\, |\tilde{\psi}(z)|^2 = \|\phi\|_2^2\, \|\psi\|_2^2 \quad ,$$

since $\|\tilde{\psi}\|_2 = \|\psi\|_2$ by the unitarity of the Fourier transformation in $L^2(\mathbb{R}^n)$. ∎

A generalization of the above result is as follows :

<u>PROPOSITION 2.8</u> : Let $2 \leq p < \infty$ and $\phi, \psi \in L^p(\mathbb{R}^n)$. Let A be as above. Then A is compact as an operator in $L^2(\mathbb{R}^n)$, and

$$\|A\| \leq (2\pi)^{\frac{n}{2} - \frac{n}{p}}\, \|\phi\|_p\, \|\psi\|_p \;.$$

<u>Remark 2.3</u> : The Fourier transform $\tilde{\psi}$ of a function ψ in $L^p(\mathbb{R}^n)$ with $p > 2$ is not necessarily defined. The correct definition of the operator A in Proposition 2.7 is $A = (2\pi)^{n/2}\, \phi(Q)\, \psi(P)$, where Q and P are the n-component position and momentum operator, respectively, defined in the example following Proposition 3.4. Thus $(F\psi(P)f)(k) = \psi(k)\, \tilde{f}(k)$, which implies formally

$$[\psi(P)f](x) = (2\pi)^{-\frac{n}{2}} \int dx\, \tilde{\psi}(y-x)\, f(y).$$

<u>Proof</u> : The proof is based on the Hölder inequality and the Hausdorff-Young inequality which are as follows :

Hölder : $$\|fg\|_r \leq \|f\|_s\, \|g\|_q \,,\quad r^{-1} = s^{-1} + q^{-1} \,. \tag{2.10}$$

Hausdorff-Young : $$\|\tilde{f}\|_q \leq (2\pi)^{\frac{n}{2} - \frac{n}{r}}\, \|f\|_r \quad , \tag{2.11}$$

with $1 \le r \le 2$, $q^{-1} + r^{-1} = 1$.

i) Let $h = (2\pi)^{n/2}\, \psi(P)\, f$. Then by the Hölder inequality, $\tilde{h} \in L^r$ with $r^{-1} = p^{-1} + \frac{1}{2}$, and

$$\|\tilde{h}\|_r \le (2\pi)^{n/2}\, \|\psi\|_p\, \|f\|_2 \quad . \text{ Notice that } 1 \le r \le 2.$$

By the Hausdorff-Young inequality, $h \in L^q$ with $q^{-1} = 1 - r^{-1} = \frac{1}{2} - p^{-1}$, and

$$\|h\|_q \le (2\pi)^{\frac{n}{2}-\frac{n}{r}}\, \|\tilde{h}\|_r \le (2\pi)^{n-\frac{n}{r}}\, \|\psi\|_p\, \|f\|_2 \quad .$$

By the Hölder inequality, $\phi h \in L^2$, since $q^{-1} + p^{-1} = \frac{1}{2}$, and

$$\|\phi h\|_2 \le \|\phi\|_p\, \|h\|_q \le (2\pi)^{n-\frac{n}{r}}\, \|\phi\|_p\, \|\psi\|_p\, \|f\|_2 \, .$$

Hence A is a bounded operator and

$$\|A\| \le (2\pi)^{\frac{n}{2}-\frac{n}{p}}\, \|\phi\|_p\, \|\psi\|_p \quad . \tag{2.12}$$

ii) Let χ_R be the characteristic function of the ball $S_R = \{x \mid |x| \le R\}$, i.e. $\chi_R(x) = 1$ if $|x| \le R$, $\chi_R(x) = 0$ if $|x| > R$. Then

$$\phi\chi_R \in L^2(\mathbb{R}^n) \cap L^p(\mathbb{R}^n)$$

$$\psi\chi_R \in L^2(\mathbb{R}^n) \cap L^p(\mathbb{R}^n)$$

(in fact by the Hölder inequality, we have $\|\phi\chi_R\|_2 \le \|\phi\|_p\, \|\chi_R\|_r$ with $r^{-1} = \frac{1}{2} - p^{-1}$, and $\|\chi_R\|_r^r$ is the volume of $S_R < \infty$).

Furthermore $\|\phi(1 - \chi_R)\|_p \to 0$ as $R \to \infty$, $\|\psi(1 - \chi_R)\|_p \to 0$ as $R \to \infty$, since $\phi, \psi \in L^p$.

By Proposition 2.7, the integral operator A_R with kernel $(\phi\chi_R)(x)\ (\widetilde{\psi\chi_R})(y-x)$ is Hilbert-Schmidt, hence compact. Now $A - A_R$ is the integral operator with kernel

$$\phi(x)\ \tilde{\psi}(y-x) - (\phi\chi_R)(x)(\widetilde{\psi\chi_R})(y-x)$$

$$= \phi(x)\ [(\tilde{\psi} - \widetilde{\psi\chi_R})(y-x)] + [(\phi - \phi\chi_R)(x)]\ (\widetilde{\psi\chi_R})(y-x),$$

so that by (2.12)

$$\|A - A_R\| \le \|\phi\|_p\ \|\psi(1 - \chi_R)\|_p + \|\phi(1 - \chi_R)\|_p\ \|\psi\|_p \to 0 \text{ as } R \to \infty .$$

Hence A is the uniform limit of the sequence $\{A_R\}_{R=1}^{\infty}$ of compact operators, i.e. A is compact by Lemma 2.4. ∎

We now say a few words about unbounded operators.

<u>DEFINITION</u> : A linear operator B is an *extension* of A if $D(B) \supseteq D(A)$ and $Bf = Af$ for each $f \in D(A)$. One also says that A is the *restriction* of B to D(A) and writes $A \subseteq B$.

<u>DEFINITION</u> : A linear operator A is *closable* if the following holds : whenever $\{f_n\} \in D(A)$, $\underset{n\to\infty}{\text{s-lim}}\ f_n = 0$ and $\underset{n\to\infty}{\text{s-lim}}\ Af_n$ exists, then $\underset{n\to\infty}{\text{s-lim}}\ Af_n = 0$.

If A is closable, one may define an extension of A, its *closure* $\bar{A}$, as follows :

$D(\bar{A}) = \{f \in \mathcal{H} \mid \exists$ a sequence $\{f_n\} \in D(A)$ with $\underset{n\to\infty}{\text{s-lim}}\ f_n = f$ and $\{Af_n\}$ strongly Cauchy$\}$. If $f \in D(\bar{A})$, set $\bar{A}f = \text{s-lim}\ Af_n$, where f_n is as above.

Consistency : If $\{f_n\}$, $\{f'_n\}$ are two sequences converging to f, then $\|f_n - f'_n\| \to 0$ as $n \to \infty$. If $\{Af_n\}$ and $\{Af'_n\}$ are both Cauchy, then $\|A(f_n - f'_n)\| \to 0$ as $n \to \infty$, since A is closable. Hence $\text{s-lim}_{n\to\infty} Af_n = \text{s-lim}_{n\to\infty} Af'_n$, so that $\bar{A}f$ is well defined.

DEFINITION : A is *closed* if $A = \bar{A}$.

Example 2.3 : If $A \in \mathcal{B}(\mathcal{H})$, then A is closed. In fact, let $f \in \mathcal{H}$, $\|f_n - f\| \to 0$ as $n \to \infty$. Then $\|Af_n - Af\| \le \|A\|\,\|f_n - f\| \to 0$, so that Af_n converges strongly to Af. Hence $\bar{A}f = Af$ for each $f \in \mathcal{H}$, i.e. $\bar{A} = A$.

Example 2.4 : Let $\{e_i\}$ be an orthonormal basis of $\mathcal{H}$, $\dim \mathcal{H} = \infty$, and define A as follows :

D(A) is the set of all finite linear combinations of the vectors $e_1, e_2, \ldots$, and $A(\Sigma\, \alpha_k e_k) = \Sigma\, k\, \alpha_k\, e_1$. A is not closable.

For the proof, consider the sequence $f_n = \frac{1}{n} e_n$. Clearly $\text{s-lim}_{n\to\infty} f_n = 0$, since $\|f_n\| = \frac{1}{n}\|e_n\| = \frac{1}{n}$. One has $Af_n = \frac{1}{n} Ae_n = \frac{1}{n} ne_1 = e_1$, hence $\{Af_n\}$ is strongly Cauchy, $\text{s-lim}_{n\to\infty} Af_n = e_1$. Since $e_1 \neq 0$, $\text{s-lim}_{n\to\infty} Af_n \neq 0$, and A is not closable.

DEFINITION : Let A be densely defined in $\mathcal{H}$. The *adjoint* A^* of A is the following linear operator :

$$g \in D(A^*) \Leftrightarrow \exists g^* \in \mathcal{H} \text{ such that } (g,Af) = (g^*,f) \text{ for all } f \in D(A) ,$$

and if $g \in D(A^*)$, then $A^*g = g^*$.

Consistency : One must show that g^* is unique. If g_1^* is such that $(g,Af) = (g_1^*,f)$ for $f \in D(A)$, then $(g^* - g_1^*,f) = 0$ for all $f \in D(A)$. Since D(A) is assumed to be dense in $\mathcal{H}$, $g^* - g_1^*$ is orthogonal to a dense set, and $g^* - g_1^* = 0$. Hence $g_1^* = g^*$.

LEMMA 2.9 : Let $D(A)$ be dense in $\mathcal{H}$.

(a) If $A \subseteq B$, then $B^* \subseteq A^*$.

(b) A^* is closed.

(c) If A is closable, then $(\bar{A})^* = A^*$.

Proof : We prove (b) and leave (a) and (c) as an exercise.

(b) Let $g_n \in D(A^*)$, $\text{s-}\lim_{n\to\infty} g_n = g$ and $\text{s-}\lim_{n\to\infty} A^* g_n = h$. Let $f \in D(A)$. Then :

$$(g\,, Af) = \lim_{n\to\infty} (g_n\,, Af) = \lim_{n\to\infty} (A^* g_n, f) = (h\,,f) \ . \qquad (2.13)$$

First, let $g = 0$. Then $(h\,,f) = 0$ for all $f \in D(A)$, which implies $h = 0$, since $D(A)$ is dense. This means that A^* is closable.

Now let g be arbitrary. (2.13) then says that $g \in D(A^*)$ and $A^* g = h$. In view of the definition of the closure, this implies that $D(\overline{A^*}) \subseteq D(A^*)$ and $\overline{A^*}g = A^* g \quad \forall g \in D(\overline{A^*})$, which proves (b). ∎

DEFINITION : Let A be closed. The complex number z belongs to the *resolvent set* $\rho(A)$ of A if $(A - z)$ is invertible and $(A - z)^{-1} \in \mathcal{B}(\mathcal{H})$. The operator-valued function :

$$z \longmapsto (A - z)^{-1} \quad \text{from} \quad \rho(A) \quad \text{to} \quad \mathcal{B}(\mathcal{H}) \quad \text{is called the } \textit{resolvent} \text{ of } A.$$

Remark 2.4 : $z \in \rho(A)$ if and only if $(A - z)$ is injective and surjective. In fact :

(a) $(A - z)^{-1}$ exists if and only if $(A - z)$ is injective.

(b) if $(A - z)^{-1} \in \mathcal{B}(\mathcal{H})$, then clearly $(A - z)$ must be surjective.

(c) if $(A-z)$ is surjective, $(A-z)^{-1}$ is defined everywhere. Also $(A-z)^{-1}$ is closed. Then $(A-z)^{-1}$ is bounded by the closed graph theorem, i.e. $(A-z)^{-1} \in \mathcal{B}(\mathcal{H})$.

DEFINITION : Let A be closed. The *spectrum* $\sigma(A)$ of A is the complement of $\rho(A)$ in the complex plane :

$$\sigma(A) = \mathbb{C} \setminus \rho(A) \quad .$$

The main purpose of this course is to study the spectrum of Schrödinger operators.

DEFINITION : The linear operator F is a *projection* (more precisely an *orthogonal projection*) if $D(F) = \mathcal{H}$ and $F = F^2 = F^*$.

The range of a projection is a (closed) subspace M (the set of vectors verifying $f = Ff$). Conversely, given a subspace M, there exists a projection F whose range is M.

If fact, let $f \in \mathcal{H}$. If we define by $Ff = f_1$, where f_1 is as in the projection theorem, then F is a projection and its range is M.

3 Self-adjointness

DEFINITION : Let $D(A)$ be dense in $\mathcal{H}$.

(i) A is *symmetric* if A^* is an extension of A, i.e. $A \subseteq A^*$.

(ii) A is *self-adjoint* if $A = A^*$.

(iii) A is *essentially self-adjoint* if $\bar{A} = A^*$.

Remarks :

(1) A is symmetric if $(f, Ag) = (Af, g)$ for all $f,g \in D(A)$.

(2) By Lemma 2.9(c), $A^* = \bar{A}^*$. Hence A is essentially self-adjoint if and only if $\bar{A} = \bar{A}^*$, i.e. if and only if the closure $\bar{A}$ of A is self-adjoint.

(3) Each self-adjoint operator is closed (by Lemma 2.9(b)).

(4) Each self-adjoint and each essentially self-adjoint operator is symmetric.

(5) The closure of a symmetric operator need not be self-adjoint. A symmetric operator may have self-adjoint extensions or not. For our purpose the important case is that where a symmetric operator has exactly one self-adjoint extension, i.e. where it is essentially self-adjoint.

PROPOSITION 3.1 : (*Basic criterion for self-adjointness*)

The symmetric operator A in $\mathcal{H}$ is self-adjoint if and only if both of the operators $A \pm i$ have range $\mathcal{H}$:

$$\{f \mid f = (A+i)g,\ g \in D(A)\} = \mathcal{H} = \{f \mid f = (A-i)g,\ g \in D(A)\} .$$

LEMMA 3.2 : Let A be symmetric. A is closable, and the ranges of $\bar{A} \pm i$ are the closure of the ranges of $A \pm i$ respectively.

Proof of Lemma 3.2 :

(i) Since A is symmetric, A has a closed extension, namely A^*. Suppose $\{f_n\} \in D(A)$, $\text{s-lim}_{n\to\infty} f_n = 0$ and $\text{s-lim}_{n\to\infty} Af_n = h$. Since $A \subseteq A^*$, $Af_n = A^*f_n$, so that $\text{s-lim}_{n\to\infty} A^*f_n = h$. Since A^* is closed (Lemma 2.9(b)), this means that $h = A^*(\text{s-lim}_{n\to\infty} f_n) = A^*0 = 0$. Thus A is closable [by the definition of a closable operator].

(ii) Let $f \in D(\bar{A})$. There exists $\{f_n\} \in D(A)$ with $f = \text{s-lim}_{n\to\infty} f_n$ and $\bar{A}f = \text{s-lim}_{n\to\infty} Af_n$. Thus $\text{s-lim}_{n\to\infty} (A \pm i)f_n = (\bar{A} \pm i)f$, and we obtain

$$(\bar{A} \pm i)\, D(\bar{A}) \subseteq \overline{\text{range}\ (A \pm i)}\ . \tag{3.1}$$

[An equation bearing two signs is valid separately for the upper and for the lower sign].

(iii) Since A is symmetric, we have for $f \in D(A)$

$$\left.\begin{aligned} \|(A \pm i)f\|^2 &= ((A \pm i)f, (A \pm i)f) = \|Af\|^2 + \|f\|^2 \pm i(Af\,, f) \\ &\mp i(f\,, Af) = \|Af\|^2 + \|f\|^2\ . \end{aligned}\right\} \tag{3.2}$$

Suppose e.g. that $g \in \overline{(A+i)\,D(A)}$. There exists $\{f_n\} \in D(A)$ such that $\text{s-lim}_{n\to\infty} (A+i)f_n = g$. By (3.2)

$$\|f_n - f_m\| \le \|(A+i)(f_n - f_m)\| \to 0 \quad \text{as} \quad n,m \to \infty,$$

$$\|Af_n - Af_m\| \le \|(A+i)(f_n - f_m)\| \to 0 \quad \text{as} \quad n,m \to \infty,$$

since $(A+i)f_n$ is strongly Cauchy. By the definition of the closure,

$h \equiv \text{s-}\lim_{n\to\infty} f_n \in D(\overline{A + i}) \equiv D(\bar{A})$ and $g = (\bar{A} + i)h$. Thus $g \in \text{range}\,(\bar{A} + i)$, which shows that the equality sign holds in (3.1). ∎

Proof of Proposition 3.1 :

(i) Suppose $(A \pm i)D(A) = \mathcal{H}$. Let $g \in D(A^*)$. There exists $h \in D(A)$ such that $(A^* - i)g = (A - i)h$. We have used $(A - i)D(A) = \mathcal{H}$. Since $A \subseteq A^*$ one has $(A - i)h = (A^* - i)h$, leading to $(A^* - i)(g - h) = 0$. Now for $f \in D(A)$

$$0 = ((A^* - i)(g - h),f) = ((g - h),(A + i)f) \quad .$$

Since $(A + i)D(A) = \mathcal{H}$, we have $g - h = 0$, i.e. $g \in D(A)$. Thus $D(A^*) \subseteq D(A)$. Since $A \subseteq A^*$, we must have $A = A^*$.

(ii) Suppose $A = A^*$. By Lemma 3.2 [and (3) of the Remark], $(A \pm i)D(A)$ are closed subspaces of $\mathcal{H}$. If for instance $(A + i)D(A) \neq \mathcal{H}$, let $g \in \{(A + i)D(A)\}^{\perp}$. It follows that $g \in D((A + i)^*) = D(A^* - i) = D(A^*)$ and $(A^* - i)g = 0$ by using the definition of the adjoint. Thus

$$(g,g) = (g,-iA^*g) = (iAg,g) = (iA^*g,g) = (-g,g) = -(g,g) \quad ,$$

whence $(g,g) = 0$, i.e. $g = 0$. Hence $\{(A + i)D(A)\}^{\perp} = \{0\}$, i.e. $(A + i)D(A) = \mathcal{H}$. Similarly $(A - i)D(A) = \mathcal{H}$. ∎

COROLLARY 3.3 : The symmetric operator A is essentially self-adjoint if and only if the range *of both of the operators* $A \pm i$ is dense in $\mathcal{H}$ (i.e. $(A + i)D(A) = (A - i)D(A) = \mathcal{H}$).

Proof : This corollary is left as exercise.

Now consider some particular self-adjoint operators, namely maximal multiplication operators in L^2-spaces.

PROPOSITION 3.4 : Let $\mathcal{H} = L^2(\Sigma,\mu)$. Let $\psi : \Sigma \to \mathbb{R}$ be μ-measurable and finite μ-a.e. Define an operator A in $\mathcal{H}$ by

$$D(A) = \{f \in L^2(\Sigma,\mu) \mid \psi(.)f(.) \in L^2(\Sigma,\mu)\}$$

and

$$(Af)(x) = \psi(x)\, f(x) \quad \text{for} \quad f \in D(A) .$$

(a) A is self-adjoint.

(b) $A \in \mathcal{B}(\mathcal{H})$ if and only if ψ is essentially bounded, in which case $\|A\| = \operatorname*{ess\,sup}_{x\in\Sigma} |\psi(x)| = \|\psi\|_\infty$.

Proof ·

(i) Clearly D(A) is a linear manifold. In addition it is dense in $\mathcal{H}$:

Set $$\Delta_m = \{x \in \Sigma \mid |\psi(x)| \le m\} , \quad m = 1,2,\ldots \quad .$$

Then $L^2(\Delta_m,\mu) \subseteq D(A)$, since

$$\int_{\Delta_m} |\psi(x)|^2 |f(x)|^2 \, d\mu(x) \le m^2 \int_{\Delta_m} |f(x)|^2 \, d\mu(x) \le m^2 \|f\|^2 < \infty. \qquad (3.3)$$

Since ψ is finite μ-a.e., $\bigcup_{m=1}^{\infty} L^2(\Delta_m,\mu)$ is dense in $\mathcal{H}$, i.e. D(A) is dense.

(ii) Since ψ is real-valued, A is symmetric : $(f,Ah) = (Af,h)$ for all $f,h \in D(A)$. Let $h \in \mathcal{H}$ and define

$$h_\pm(x) = \frac{h(x)}{\psi(x) \pm i} \, .$$

Now $|\psi(x) \pm i|^{-1} \le 1$ and $|\psi(x)|\, |\psi(x) \pm i|^{-1} \le 1$. Thus as in (3.3) :

$$\int_{\Sigma} |h_{\pm}(x)|^2 \, d\mu(x) \le \int_{\Sigma} |h(x)|^2 \, d\mu(x) ,$$

$$\int_{\Sigma} |\psi(x)|^2 \, |h_{\pm}(x)|^2 \, d\mu(x) \le \int_{\Sigma} |h(x)|^2 \, d\mu(x) ,$$

so that $h_{\pm} \in L^2(\Sigma,\mu)$ and $h_{\pm} \in D(A)$. Since h is arbitrary, this means range $(A \pm i) = \mathcal{H}$, i.e. $A = A^*$ by Proposition 3.1. This proves (a).

(iii) Let $M \equiv \operatorname{ess\,sup}_{x \in \Sigma} |\psi(x)| < \infty$. As in (3.3) $\|Af\|^2 \le M^2 \|f\|^2$, i.e. $\|A\| \le M$.

If $M_0 < M$, set $\Delta_0 = \{x \in \Sigma \mid |\psi(x)| > M_0\}$. One has $\mu(\Delta_0) > 0$. If $f \in L^2(\Delta_0,\mu)$, then $\|Af\|^2 \ge M_0^2 \|f\|^2$, i.e. $\|A\| > M_0$. Hence $\|A\| = M$.

If ψ is not essentially bounded, one finds similarly that $\|A\| \ge m$ for any finite m, i.e. A cannot be bounded. This proves (b).■

Examples : Let $\mathcal{H} = L^2(\mathbb{R}^n)$, with Lebesgue measure.

3.1. Let $\psi(x) = x_j$ and denote by Q_j the corresponding maximal multiplication operator, defined by

$$(Q_j f)(x) = x_j f(x) \quad , \quad f \in D(Q_j) \quad .$$

Q_j is called the j-th component of the *position operator* in physics. The position operator will be denoted by $Q = (Q_1,\ldots,Q_n)$. We also set $|Q| = (\sum_{i=1}^{n} Q_i^2)^{1/2}$. The maximal multiplication operator by a function $\psi(x)$ will be denoted by $\psi(Q)$.

3.2. The operator $-i \frac{\partial}{\partial x_j}$ is naturally defined on C_0^1-functions. It is essentially self-adjoint but not closed on $C_0^1(\mathbb{R}^n) \subset L^2(\mathbb{R}^n)$. One notices that

$$[\mathcal{F}(-i \frac{\partial}{\partial x_j} f)](k) = k_j \tilde{f}(k) \ .$$

Hence $-i \frac{\partial}{\partial x_j}$ has a self-adjoint extension P_j, the maximal multiplication operator by k_j on the Fourier transforms of $L^2(\mathbb{R}^n)$:

$$D(P_j) = \{f \in L^2(\mathbb{R}^n) \mid \int k_j^2 \, |\tilde{f}(k)|^2 \, dk < \infty\}$$

and

$$(\widetilde{P_j f})(k) = k_j \tilde{f}(k) \ , \quad f \in D(P_j) \ .$$

P_j is called the j-th component of the *momentum operator*.

3.3. Similarly $-\Delta$ is defined naturally on $C_o^2(\mathbb{R}^n) \subset L^2(\mathbb{R}^n)$. It can be shown that it is essentially self-adjoint. To obtain its self-adjoint closure K_o, one defines as above

$$D(K_o) = \{f \in L^2(\mathbb{R}^n) \mid \int [\sum_{j=1}^{n} k_j^2]^2 \, |\tilde{f}(k)|^2 \, dk < \infty\}$$

and

$$(\widetilde{K_o f})(k) = \sum_{j=1}^{n} k_j^2 \, \tilde{f}(k) \quad , \quad f \in D(K_o) \ .$$

<u>LEMMA 3.5</u> : Let $n = 3$. Then every $f \in D(K_o)$ is equivalent to a bounded, uniformly continuous function of x.

<u>Remark</u> : The result is not true for $n > 3$. The case $n = 3$ is the physically interesting one which we shall treat in more detail.

Proof : Let $a > 0$. Then

$$\int_{\mathbb{R}^3} (k^2 + a^2)^{-2}\, dk = 4\pi \int_0^\infty (k^2 + a^2)^{-2} k^2 dk = \frac{\pi^2}{a} .$$

From the Schwarz inequality in $L^2(\mathbb{R}^3)$ we get :

$$\left.\begin{aligned} \left[\int |\tilde{f}(k)|\, dk\right]^2 &= \left[\int \{(k^2 + a^2)\tilde{f}(k)\}(k^2 + a^2)^{-1}\, dk\right]^2 \\ &\le \frac{\pi^2}{a} \int |(k^2 + a^2)\, \tilde{f}(k)|^2\, dk = \frac{\pi^2}{a} \|(K_0 + a^2)f\|^2 \quad , \end{aligned}\right\} \tag{3.4}$$

therefore $\tilde{f} \in L^1(\mathbb{R}^3)$. We have

$$\begin{aligned} |f(x)| &= |(2\pi)^{-\frac{3}{2}} \int e^{ik.x}\, \tilde{f}(k)\, dk| \\ &\le (2\pi)^{-\frac{3}{2}} \int |\tilde{f}(k)|\, dk \le c\, a^{-\frac{1}{2}} \|(K_0 + a^2)f\|, \end{aligned}$$

which implies that f is bounded.
The uniform continuity of f will not be used, and we omit its proof (Problem 3.2). ∎

We now give a theorem from perturbation theory. First, assume that $A = A^*$ and $B = B^*$ and that $D(A) \cap D(B)$ is dense. Define $A + B$ by $D(A+B) = D(A) \cap D(B)$ and $(A+B)f = Af + Bf$. Since for $f,g \in D(A+B)$

$$\begin{aligned} (f,(A + B)g) &= (f,Ag + Bg) = (f,Ag) + (f,Bg) \\ &= (Af,g) + (Bf,g) = ((A+B)f,g) \ , \end{aligned}$$

$A + B$ is symmetric. If B can be considered as a perturbation of A, one can prove that $A + B$ is self-adjoint.

We first give some preliminary definitions.

DEFINITION : Let A,B be linear operators in $\mathcal{H}$. If $D(A) \subseteq D(B)$ and there exist $\alpha,\beta \geq 0$ such that for all $f \in D(A)$

$$\|Bf\| \leq \alpha\|f\| + \|Af\| , \tag{3.5}$$

we say that B is *A-bounded*. The greatest lower bound β_0 of all β for which (3.5) holds is called the *A-bound* of B.

Example 3.4 : If $B \in \mathcal{B}(\mathcal{H})$, then $\|Bf\| \leq \|B\| \, \|f\|$. B is bounded, and it is A-bounded with A-bound 0.

Example 3.5 : Let $\mathcal{H} = L^2(\mathbb{R}^3)$, B the maximal multiplication operator by $B(x)$, where $\int |B(x)|^2 \, dx < \infty$. Let $A = K_0$. By Lemma 3.5, $f \in D(K_0)$ implies that $|f(x)| \leq c\, a^{-\frac{1}{2}} \|(K_0 + a^2)f\|$, $a > 0$. Hence

$$\|Bf\|^2 = \int |B(x)|^2 \, |f(x)|^2 \, dx \leq c^2 \, a^{-1}\|(K_0 + a^2)f\|^2 \int dx \, |B(x)|^2 < \infty.$$

Thus $D(K_0) \subseteq D(B)$. Also

$$\|Bf\| \leq c\, a^{-\frac{1}{2}} \|B\|_2 \, [a^2\|f\| + \|K_0 f\|]$$

$$= c\, a^{\frac{3}{2}} \|B\|_2\|f\| + c\, a^{-\frac{1}{2}} \|B\|_2\|K_0 f\| \equiv \alpha\|f\| + \beta\|K_0 f\| \quad ,$$

with $\beta = c\, a^{-\frac{1}{2}} \|B\|_2$.

Since $a > 0$ is arbitrary, (3.5) holds for any $\beta > 0$, i.e. B is K_0-bounded with K_0-bound 0 (but it is not possible to set $\beta = 0$ in (3.5), unless $B(.)$ is essentially bounded on $\mathbb{R}^3$!).

Example 3.6 : $\mathcal{H} = L^2(\mathbb{R}^3)$, $A = K_0$, B the multiplication operator by $B(x) = B_1(x) + B_2(x)$, $B_1(.) \in L^2(\mathbb{R}^3)$ and $B_2(.)$ essentially bounded [i.e. $B_2 \in \mathcal{B}(\mathcal{H})$ by Proposition 3.4]. Then, by Examples 3.4 and 3.5 :

$$\|Bf\| \leq \|B_1 f\| + \|B_2 f\| \leq c\, a^{-\frac{1}{2}} \|B_1\|_2 \|K_0 f\| + c\, a^{\frac{3}{2}} \|B_1(.)\|_2 \|f\| + \|B_2\| \|f\| ,$$

where $\|B_2\| \equiv [\operatorname{ess\,sup}_{x \in \mathbb{R}^3} |B_2(x)|]$.

Again B is K_0-bounded with K_0-bound 0.

PROPOSITION 3.6 : Let $A = A^*$ and B be a symmetric operator. If B is A-bounded with A-bound $\beta < 1$, then A + B is self-adjoint with $D(A + B) = D(A)$. Furthermore B is (A+B)-bounded.

COROLLARY 3.7 : Let $\mathcal{H} = L^2(\mathbb{R}^3)$. Let V be the maximal multiplication operator by $V(x) = V_1(x) + V_2(x)$ with $V_1(.) \in L^2(\mathbb{R}^3)$, $V_2(.)$ essentially bounded, and V real. Then $K_0 + V$ is self-adjoint on $D(K_0)$.

Proof : Use Proposition 3.6 and Example 3.6. ∎

Proof of Proposition 3.6 :

(i) Let $C = C^*$. Then for $f \in D(C)$

$$\left.\begin{aligned} \|(C \pm i)f\|^2 &= \|Cf\|^2 + \|f\|^2 \pm i(Cf,f) \mp i(f,Cf) \\ &= \|Cf\|^2 + \|f\|^2 \quad . \end{aligned}\right\} \qquad (3.6)$$

If $(C \pm i)f = 0$, then $\|f\| = 0$, i.e. $f = 0$. Thus $(C \pm i)^{-1}$ exists. Since $C = C^*$, $(C \pm i)D(C) = \mathcal{H}$ by Proposition 3.1. Thus $(C \pm i)^{-1}$ is defined everywhere and maps $\mathcal{H}$ onto $D(C)$.

(ii) (3.5) implies for $f \in D(A)$

$$\|Bf\|^2 \leq \alpha^2\|f\|^2 + \beta^2\|Af\|^2 + 2\alpha\beta\|f\| \, \|Af\| \, . \qquad (\beta < 1)$$

Now for $\eta > 0$

$$0 \leq (\frac{\alpha}{\eta}\|f\| - \beta\eta\|Af\|)^2 = \frac{\alpha^2}{\eta^2}\|f\|^2 + \beta^2\eta^2\|Af\|^2 - 2\alpha\beta\|f\| \, \|Af\| \, .$$

We add these two inequalities :

$$\left.\begin{aligned} \|Bf\|^2 &\leq \alpha^2(1 + \frac{1}{\eta^2}) \|f\|^2 + \beta^2(1 + \eta^2) \|Af\|^2 \\ &\equiv \alpha'^2 \|f\|^2 + \beta'^2\|Af\|^2 \, . \end{aligned}\right\} \tag{3.7}$$

Since $\beta < 1$, we may choose $\eta > 0$ such that $\beta' \equiv \beta\sqrt{1 + \eta^2} < 1$. We may also assume that $\alpha > 0$, $\beta > 0$ and set $\lambda = \beta'/\alpha'$. Notice that $\lambda > 0$.

(iii) By using (3.6) with $C = \lambda A$, (3.7) gives

$$\|\frac{B}{\alpha'} f\|^2 \leq \|f\|^2 + \|\lambda Af\|^2 = \|(\lambda A \pm i)f\|^2 \, .$$

Since $(\lambda A \pm i)^{-1}$ maps $\mathcal{H}$ onto $D(\lambda A) = D(A)$, we may set $f = (\lambda A \pm i)^{-1}g$:

$$\|\frac{B}{\alpha'} (\lambda A \pm i)^{-1}g\| \leq \|g\|$$

or

$$\|\lambda B(\lambda A \pm i)^{-1}g\| \leq \beta'\|g\| ,$$

which implies $\|\lambda B(\lambda A \pm i)^{-1}\| \leq \beta' < 1$.
Now use the Neumann series (Proposition 2.3) :

$\{I + \lambda B(\lambda A \pm i)^{-1}\}^{-1}$ exists and is in $\mathcal{B}(\mathcal{H})$, i.e. the operators $I + \lambda B(\lambda A \pm i)^{-1}$ map $\mathcal{H}$ onto $\mathcal{H}$. Now $\lambda A \pm i$ map D(A) onto $\mathcal{H}$ by Proposition 3.1. Thus :

$$[I + \lambda B(\lambda A \pm i)^{-1}]\,(\lambda A \pm i) \equiv \lambda(A + B) \pm i$$

map D(A) onto $\mathcal{H}$. Since $D(A) = D(A + B) = D(\lambda(A + B))$, $\lambda(A + B)$ is self-adjoint by Proposition 3.1. Since $\lambda \in \mathbb{R}$, $\lambda \neq 0$, A + B is self-adjoint.

(iv) From the triangle inequality :

$$\|Af\| = \|(A + B)f - Bf\| \leq \|(A + B)f\| + \|Bf\| .$$

We insert this into (3.5) :

$$\|Bf\| \leq \alpha\|f\| + \beta\|Af\| \leq \alpha\|f\| + \beta\|(A + B)f\| + \beta\|Bf\| .$$

Now :

$$(1 - \beta)\|Bf\| \leq \alpha\|f\| + \beta\|(A + B)f\|$$

$$\|Bf\| \leq \frac{\alpha}{1-\beta}\|f\| + \frac{\beta}{1-\beta}\|(A+B)f\| .$$

Hence B is (A+B)-bounded (its (A+B)-bound may be larger then 1 though!). ∎

The generalization of Corollary 3.7 to n dimensions is as follows : For $n = 1,2$, the same result can be obtained by the same method. For $n > 3$, one notices that

$$\int_{\mathbb{R}^n} dk\ (k^2 + a^2)^{-p} \equiv N_a < \infty$$

if $p > n/2$, and $N_a \to 0$ as $a \to \infty$.

The resolvent $(K_o + a^2)^{-1}$ of K_o is an integral operator in $L^2(\mathbb{R}^n)$ whose kernel is the Fourier transform (up to a factor) of $(k^2 + a^2)^{-1}$. Hence

$$\| V(K_o + a^2)^{-1} \| \le c \| V \|_p \| (k^2 + a^2)^{-1} \|_p \quad ,$$

[see Proposition 2.8]. If $V \in L^p(\mathbb{R}^n)$ for some $p > n/2$, then $\| V(K_o + a^2)^{-1} \| \equiv \beta < 1$ if a is sufficiently large. Then for $f \in D(K_o)$

$$\| Vf \| \le \| V(K_o + a^2)^{-1} \| \ \| (K_o + a^2) f \|$$

$$\le \beta \| K_o f \| + a^2 \beta \| f \| \qquad \text{with} \quad \beta < 1.$$

Hence V is K_o-bounded, with K_o-bound $\beta < 1$. Thus we have :

PROPOSITION 3.8 : If $n > 3$, $V = V_1 + V_2$ with $V_1(.) \in L^p(\mathbb{R}^n)$ for some $p > n/2$, $V_2(.)$ essentially bounded, then $K_o + V$ is self-adjoint.

Return to $n = 3$: If V is locally so singular that $V(.) \notin L^2_{loc}$, or if V is singular at infinity (e.g. $V(x) \to \pm\infty$ as $|x| \to \infty$), one has to use more powerful methods than perturbation theory to prove self-adjointness under suitable assumptions. We shall not go into this.

Another useful concept is that of relative compactness, which will however not be used later.

DEFINITION : Let A and B be closed. B is called A-*compact* if $D(A) \subseteq D(B)$ and $B(A - z)^{-1}$ is compact for some $z \in \rho(A)$.

Example 3.7 : In $L^2(\mathbb{R}^n)$, let $V(.) \in L^p(\mathbb{R}^n)$ with $p > n/2$, and $p \geq 2$. V is K_0-compact by Proposition 2.8.

PROPOSITION 3.9 : Let $A = A^*$ and B be A-compact. Then B is A-bounded with A-bound $\beta_0 = 0$.

4 Spectral theory

It is well known that a hermitian matrix in a finite-dimensional Hilbert space can be diagonalized by a unitary transformation. The spectral theorem for self-adjoint operators is the infinite-dimensional analogue of this property. A major difficulty in a infinite-dimensional space stems from the fact that a self-adjoint operator may have continuous spectrum and not only eigenvalues. We shall explain the spectral theorem and some of its consequences, but we shall not give a complete proof because we want to spend more time on the study of Schrödinger operators.

<u>DEFINITION</u> : A *spectral family* in $\mathcal{H}$ is a mapping $\lambda \mapsto E_\lambda$ from the real line into the set off all projections in $\mathcal{H}$ (i.e. $E_\lambda = E_\lambda^2 = E_\lambda^*$) having the following properties :

(S_a) $$E_\lambda E_\mu = E_\mu E_\lambda = E_{\min\{\lambda,\mu\}}, \quad \forall \lambda,\mu \in \mathbb{R} .$$

(S_b) $$\underset{\lambda\to-\infty}{\text{s-lim}}\, E_\lambda = 0, \quad \underset{\lambda\to+\infty}{\text{s-lim}}\, E_\lambda = I .$$

(S_c) $$\underset{\varepsilon\to+0}{\text{s-lim}}\, E_{\lambda+\varepsilon} = E_\lambda , \qquad \forall \lambda \in \mathbb{R} .$$

(S_a) says that the family $\{E_\lambda\}$ is *non-decreasing* : If M_λ is the range of E_λ and $\lambda \le \mu$, then $M_\lambda \subseteq M_\mu$. In fact, if $f \in M_\lambda$, then $E_\lambda f = f$. Now $E_\mu E_\lambda f = E_{\min\{\lambda,\mu\}} f = E_\lambda f$, which implies with the preceding identity that $E_\mu f = f$, i.e. $f \in M_\mu$.

(S_c) requires the *right-continuity* of $\{E_\lambda\}$.

A spectral family defines a *countably additive projection-valued Borel measure* on $\mathbb{R}$ called the *spectral measure* associated with $\{E_\lambda\}$:

If Δ is a half-open interval $\Delta = (\lambda,\mu]$, one defines

$$E_\Delta = E_\mu - E_\lambda \quad .$$

If Δ_1,Δ_2 are two such intervals with $\Delta_1 \cap \Delta_2 = \phi$, then from (S_a) it follows that

$$E_{\Delta_1} E_{\Delta_2} = E_{\Delta_2} E_{\Delta_1} = 0 \ .$$

(S_a) also implies that, for each $\lambda \in \mathbb{R}$, $\underset{\varepsilon \to +0}{\text{s-lim}}\, E_{\lambda-\varepsilon} \equiv E_{\lambda-0}$ exists and is a projection. Hence, if $\Delta = \{\lambda\}$ is a single point, we may define

$$E_{\{\lambda\}} = E_\lambda - E_{\lambda-0} \quad .$$

If $\Delta = [a,b]$, one sets $E_\Delta = E_{\{a\}} + E_{(a,b]}$. If $\Delta = (a,b)$, one has $\Delta = \bigcup_n (a,b - \frac{1}{n}]$ and thus sets

$$E_\Delta = \underset{n\to\infty}{\text{s-lim}}\, E_{(a,b-\frac{1}{n}]} \quad .$$

By the standard method one can extend this projection-valued measure from intervals to all Borel subsets of $\mathbb{R}$ and one has (Δ_i are Borel sets) :

$$E_{\Delta_1} E_{\Delta_2} = E_{\Delta_1 \cap \Delta_2} \quad . \tag{4.1}$$

$$\sum_{i=1}^{\infty} E_{\Delta_i} = E_{\bigcup_{i=1}^{\infty}\Delta_i} \qquad \text{if} \quad \Delta_i \cap \Delta_j = \phi,\ \forall i \neq j \ . \tag{4.2}$$

<u>LEMMA 4.1</u> : Let $f,g \in \mathcal{H}$. The function $\lambda \mapsto (f,E_\lambda g)$ is of bounded variation. If $\|f\| = 1$, the positive function $\lambda \mapsto \|E_\lambda f\|^2$ is normalized and of bounded

variation.

Proof :

(i) Let $\Delta = (a,b]$, let $a = \lambda_0 < \lambda_1 < \ldots < \lambda_n = b$ be a partition of Δ into intervals $\Delta_i = (\lambda_{i-1}, \lambda_i]$. Then

$$\sum_{i=1}^{n} |(f,E_{\lambda_i} g) - (f,E_{\lambda_{i-1}} g)| = \sum_{i=1}^{n} |(f,E_{\Delta_i} g)| .$$

By using the fact that $E_\Delta = E_\Delta^2 = E_\Delta^*$ and Schwarz inequality, we get

$$\begin{aligned}
\sum_{i=1}^{n} |(E_{\Delta_i} f, E_{\Delta_i} g)| &\le \sum_{i=1}^{n} \|E_{\Delta_i} f\| \ \|E_{\Delta_i} g\| \\
&\le \Big(\sum_{i=1}^{n} \|E_{\Delta_i} f\|^2\Big)^{\frac{1}{2}} \Big(\sum_{i=1}^{n} \|E_{\Delta_i} g\|^2\Big)^{\frac{1}{2}} \\
&= \Big[\sum_{i=1}^{n} (f,E_{\Delta_i} f)\Big]^{\frac{1}{2}} \Big[\sum_{i=1}^{n} (g,E_{\Delta_i} g)\Big]^{\frac{1}{2}} \qquad (4.3) \\
&= (f,E_\Delta f)^{\frac{1}{2}} (g,E_\Delta g)^{\frac{1}{2}} \\
&= (E_\Delta f,E_\Delta f)^{\frac{1}{2}} (E_\Delta g,E_\Delta g)^{\frac{1}{2}} = \|E_\Delta f\| \|E_\Delta g\| \le \|f\| \ \|g\| ,
\end{aligned}$$

since $\|E_\Delta\| \le 1$.

Now the total variation of $(f,E_\lambda g)$ over any interval Δ does not exceed $\|f\| \ \|g\|$, i.e. $\lambda \mapsto (f,E_\lambda g)$ is of bounded variation.

(ii) Set $f = g$. Then $(f,E_\lambda f) = \|E_\lambda f\|^2$ is of bounded variation. Also $\lim_{\lambda \to -\infty} \|E_\lambda f\|^2 = 0$, $\lim_{\lambda \to +\infty} \|E_\lambda f\|^2 = \|f\|^2 = 1$ by (S_b). Therefore $\lambda \mapsto \|E_\lambda f\|^2$ is normalized. ■

By virtue of Lemma 4.1, one can define *Riemann-Stieltjes integrals* with respect to the measure generated by the function $(f,E_\lambda g)$. If $\phi : \mathbb{R} \to \mathbb{C}$ is continuous, then

$$\int_a^b \phi(\lambda)\ d(f,E_\lambda g)$$

exists for each finite a,b and $f,g \in \mathcal{H}$, as a limit of Riemann-Stieltjes sums. Improper integrals are defined as

$$\int_{-\infty}^{\infty} \phi(\lambda)\ d(f,E_\lambda g) = \lim_{\substack{b \to +\infty \\ a \to -\infty}} \int_a^b \phi(\lambda)\ d(f,E_\lambda g)$$

whenever these limits exist.

PROPOSITION 4.2 : Let $\{E_\lambda\}$ be a spectral family, and define

$$D(A) = \{g \in \mathcal{H} \mid \int_{-\infty}^{+\infty} \lambda^2\ d(g,E_\lambda g) < \infty\}.$$

Then there exists a self-adjoint operator A with domain D(A) such that for all $f \in \mathcal{H}$, $g \in D(A)$:

$$(Ag,f) = \int_{-\infty}^{+\infty} \lambda\ d(E_\lambda g,f)\ .$$

One writes formally $A = \int_{-\infty}^{+\infty} \lambda\ dE_\lambda$.

Proof :

(i) D(A) *is dense* in $\mathcal{H}$: If $g \in \mathcal{H}$, set $g_n = (E_n - E_{-n})g$, $n = 1,2,\ldots$. Then $\|g-g_n\| \to 0$ by (S_b), and by Lemma 4.1

$$\int_{-\infty}^{\infty} \lambda^2 \, d(g_n, E_\lambda g_n) = \int_{-n}^{n} \lambda^2 \, d(g, E_\lambda g) \le n^2 \int_{-\infty}^{\infty} d(g, E_\lambda g) = n^2 \|g\|^2 < \infty,$$

i.e. $g_n \in D(A)$.

(ii) Fix $g \in D(A)$. Then $f \mapsto \Phi_g(f) \equiv \int_{-\infty}^{\infty} \lambda \, d(E_\lambda g, f)$ defines a bounded linear functional on $\mathcal{H}$:

$$|\Phi_g(f)| \le \|f\| \left\{\int_{-\infty}^{\infty} \lambda^2 \, d(g, E_\lambda g)\right\}^{\frac{1}{2}} \equiv M_g \|f\| \ ,$$

which follows as in the proof of Lemma 4.1. Thus $\Phi_g \in \mathcal{H}^*$ (the dual of $\mathcal{H}$). Now a Hilbert space is self-dual (this is called the theorem of Riesz), so that there exists $g^* \in \mathcal{H}$ such that

$$\Phi_g(f) = (g^*, f) \ .$$

One defines A by $Ag = g^*$. Thus

$$(Ag, f) = \int_{-\infty}^{+\infty} \lambda \, d(E_\lambda g, f) \ .$$

If also $f \in D(A)$, then

$$(g, Af) = \overline{(Af, g)} = \int_{-\infty}^{\infty} \lambda \, \overline{d(f, E_\lambda g)} = \int_{-\infty}^{\infty} \lambda \, d(E_\lambda g, f) = (Ag, f),$$

i.e. A *is symmetric*.

(iii) Let $\phi_\pm : \mathbb{R} \to \mathbb{C}$ be defined by $\phi_\pm(\lambda) = (\lambda \pm i)^{-1}$. As above, we get for all $f, g \in \mathcal{H}$

$$\int_{-\infty}^{\infty} \overline{\phi_\pm(\lambda)}\ d(g,E_\lambda f) \le \|f\| \ \{\int_{-\infty}^{\infty} |\phi_\pm(\lambda)|^2\ d(g,E_\lambda g)\}^{\frac{1}{2}} \le \|f\|\|g\| \ ,$$

since $|\phi_\pm(\lambda)| \le 1$. Thus there exist two linear operators $\phi_\pm(A)$ with domain $\mathcal{H}$ such that

$$(\phi_\pm(A)g,f) = \int_{-\infty}^{\infty} \overline{\phi_\pm(\lambda)}\ d(E_\lambda g,f) \ .$$

By using $E_\lambda E_\mu = E_{\min\{\lambda,\mu\}}$ we get

$$\int_{-\infty}^{\infty} \lambda^2\ d(\phi_\pm(A)g,E_\lambda\phi_\pm(A)g)$$

$$= \int_{-\infty}^{\infty} \lambda^2\ d_\lambda \int_{-\infty}^{+\infty} \overline{\phi_\pm(\mu)}\ d_\mu(g,E_\mu E_\lambda\phi_\pm(A)g)$$

$$= \int_{-\infty}^{\infty} \lambda^2\ d_\lambda \int_{-\infty}^{\lambda} \overline{\phi_\pm(\mu)}\ d_\mu(g,E_\mu\phi_\pm(A)g)$$

$$= \int_{-\infty}^{\infty} \lambda^2\ \overline{\phi_\pm(\lambda)}\ d_\lambda(g,E_\lambda\phi_\pm(A)g)$$

$$= \int_{-\infty}^{+\infty} \lambda^2\ \overline{\phi_\pm(\lambda)}\phi_\pm(\lambda)\ d_\lambda(g,E_\lambda g) \le \int_{-\infty}^{\infty} d(g,E_\lambda g) = \|g\|^2 \ ,$$

since $\lambda^2|\phi_\pm(\lambda)|^2 \le 1$. Thus $\phi_\pm(A)g \in D(A)$ for each $g \in \mathcal{H}$. A similar calculation gives, since $(\lambda\pm i)\ \phi_\pm(\lambda) \equiv 1$:

$$((A\pm i)\ \phi_\pm(A)g,f) = \int_{-\infty}^{\infty} \overline{(\lambda\pm i)}\ \overline{\phi_\pm(\lambda)}\ d(g,E_\lambda f) = \int_{-\infty}^{\infty} d(g,E_\lambda f) = (g,f).$$

Hence $(A\pm i)\ \phi_\pm(A)g = g$, i.e. each $g \in \mathcal{H}$ belongs to range $(A\pm i)$. Therefore A is self-adjoint by Proposition 3.1. ■

LEMMA 4.3 : Let $\{E_\lambda\}$, A be as above. Then (a) $f \neq 0$ and $(A-\mu)f = 0$ if and only if $\|E_\lambda f\|^2$ is constant except at $\lambda = \mu$. (b) f is an eigenvector of A with eigenvalue μ if and only if $E_{\{\mu\}}f = f$.

Proof : This is left as exercise.

Proposition 4.2 associates with each spectral family a self-adjoint operator A. The spectral theorem is the converse of this.

Spectral Theorem : Let A *be self-adjoint. Then there exists a unique spectral family* $\{E_\lambda\}$ *such that* A *is represented as in Proposition 4.2. Each* E_λ *commutes with* A *and with all bounded operators* B *that commute with* A, [K].

Comments :

A : One proof of the spectral theorem is essentially as follows :

(a) *Square root* : Let B be positive [i.e. $(f,Bf) \geq 0$ for all $f \in D(B)$] and self-adjoint. Then there exists a unique positive self-adjoint operator C such that $C^2 = B$. One writes $C = B^{1/2}$. [The proof of this is relatively easy if $B \in \mathcal{B}(\mathcal{H})$ but difficult if B is unbounded].

(b) *Polar decomposition* : Let B be closed and densely defined. Then $B = UG$ where G is positive self-adjoint and U is a unitary operator from $\overline{G\mathcal{H}}$ onto $\overline{B\mathcal{H}}$ [This is the analog of the representation $z = e^{i\phi}|z|$ for a complex number z].

U and G are as follows : B^*B is positive self-adjoint. Take $G = (B^*B)^{\frac{1}{2}}$ and $U : (B^*B)^{1/2} f \mapsto Bf$.

(c) Let $B_\lambda \equiv A - \lambda$, and consider its polar decomposition $B_\lambda = U_\lambda G_\lambda$.

Define $E_\lambda = I - \frac{1}{2}[U_\lambda + U_\lambda^2]$. Each E_λ is a projection, $\{E_\lambda\}$ is a spectral family and A the associated self-adjoint operator.

Henceforth we call $\{E_\lambda\}$ *the spectral family of* A.

B. If A is unbounded and $B \in \mathcal{B}(\mathcal{H})$, we say that B commutes with A if $BA \subseteq AB$, i.e. whenever $f \in D(A)$, then $Bf \in D(A)$ and $ABf = BAf$.

The second part of the spectral theorem is shown by verifying that, if $B \in \mathcal{B}(\mathcal{H})$ commutes with A, then $U_\lambda B = BU_\lambda$, implying $B_\lambda E_\lambda = BE_\lambda$. The fact that $E_\lambda A \subseteq AE_\lambda$ follows from Proposition 4.2 and using $E_\lambda E_\mu = E_{\min\{\lambda,\mu\}}$: If $g \in D(A)$, then for $\mu \in \mathbb{R}$

$$\int_{-\infty}^{\infty} \lambda^2 \, d(E_\mu g, E_\lambda E_\mu g) = \int_{-\infty}^{\mu} \lambda^2 \, d(E_\lambda g, g) < \infty,$$

i.e. $E_\mu g \in D(A)$. Similarly $(AE_\mu g, f) = (Ag, E_\mu f) = (E_\mu Ag, f)$.

Examples of spectral families will be given a little later. We now say a few words about the *functional calculus*.

If $\phi : \mathbb{R} \to \mathbb{C}$ is a continuous function, we may define as in Proposition 4.2 a linear operator, denoted $\phi(A)$, as follows

$$D(\phi(A)) = \{g \in \mathcal{H} \mid \int_{-\infty}^{\infty} |\phi(\lambda)|^2 \, d(g, E_\lambda g) < \infty ,$$

and for $g \in D(A)$, $f \in \mathcal{H}$, there exists $g^* \in \mathcal{H}$ such that

$$(\phi(A)g, f) \equiv (g^*, f) = \int_{-\infty}^{+\infty} \overline{\phi(\lambda)} \, d(g, E_\lambda f) .$$

[Note that $g^* = \phi(A)g$]. $D(\phi(A))$ is dense and $\|\phi(A)g\|^2 = \int_{-\infty}^{\infty} |\phi(\lambda)|^2 \, d(g, E_\lambda g)$, which is verified by a calculation similar to that in part (iii) of the proof of Proposition 4.2.

This is a natural way of defining a function of a self-adjoint operator. An other possibility would be to define $\phi(A)$ as a series for certain functions ϕ. For instance, if $A \in \mathcal{B}(\mathcal{H})$, $\exp(A) = \Sigma_{n=0}^{\infty} \frac{A^n}{n!}$, and the series converges in operator norm since $\|A\| < \infty$. If A is unbounded, it is not clear (without using the spectral theorem), whether there is a dense set $\mathcal{D}$ such that A^n is defined on $\mathcal{D}$ for all n.

PROPOSITION 4.4 : (resolvent of A) Let $z \in \mathbb{C}$, $\text{Im}z \neq 0$, and $\psi_z(\lambda) = (\lambda-z)^{-1}$. Then $\psi_z(A) = (A-z)^{-1}$,

$$\psi_z(A) \in \mathcal{B}(\mathcal{H}) \quad \text{and} \quad \|\psi_z(A)\| \le \frac{1}{|\text{Im } z|} .$$

In particular $z \in \rho(A)$, i.e. the spectrum of a self-adjoint operator is a subset of the real line.

Proof : $$\|\psi_z(A)g\|^2 = \int \frac{1}{|\lambda-z|^2} d(g,E_\lambda g) \le \frac{1}{|\text{Im } z|^2} \int d(g,E_\lambda g) .$$

Thus $\psi_z(A) \in \mathcal{B}(\mathcal{H})$, $\|\psi_z(A)\| \le |\text{Im } z|^{-1}$. The fact that $\psi_z(A) = (A-z)^{-1}$ is established as in part (iii) of the proof of Proposition 4.2, where we already studied the case $z = \pm i$. ■

If $z = \mu$ is real and $E_{(\mu-\varepsilon,\mu+\varepsilon]}g = g$, then

$$\|\psi_\mu(A)g\|^2 = \int_{\mu-\varepsilon}^{\mu+\varepsilon} \frac{1}{(\lambda-\mu)^2} d(g,E_\lambda g) \ge \frac{1}{\varepsilon^2} \|g\|^2 .$$

If $\mu \in \rho(A)$, then $\|\psi_\mu(A)\| = \|(A-\mu)^{-1}\| \le M < \infty$, i.e. there exists $\varepsilon > 0$ such that $E_{(\mu-\varepsilon,\mu+\varepsilon]} = 0$. Thus, in the integral defining $\phi(A)$, it suffices to integrate over $\sigma(A)$:

$$(\phi(A)g,f) = \int_{\sigma(A)} \overline{\phi(\lambda)}\, d(g,E_\lambda f) .$$

Conversely, if μ is a point of constancy of $\{E_\lambda\}$, i.e. if $E_{\mu-\eta} = E_{\mu+\eta}$ for some $\eta > 0$, then $\mu \in \rho(A)$. Indeed, in this case

$$\int_{-\infty}^{\infty} \frac{1}{\lambda-\mu}\, d(g,E_\lambda g) = \int_{-\infty}^{\mu-\varepsilon} \frac{1}{\lambda-\mu}\, d(g,E_\lambda g) + \int_{\mu+\varepsilon}^{\infty} \frac{1}{\lambda-\mu}\, d(g,E_\lambda g)$$

so that $(A-\mu)^{-1} \in \mathcal{B}(\mathcal{H})$ and $\|(A-\mu)^{-1}\| \le \frac{1}{\varepsilon}$. It can be shown that $\rho(A)$ is an open subset of $\mathbb{C}$, hence $\sigma(A)$ *is a closed subset of* $\mathbb{R}$.

We now give some simple facts about the functional calculus.

PROPOSITION 4.5 : Let $\phi,\phi_1,\phi_2 : \mathbb{R} \to \mathbb{C}$ be continuous bounded functions. Then

(a) $D(\phi(A)) = \mathcal{H}$, $\phi(A) \in \mathcal{B}(\mathcal{H})$ and

$$\|\phi(A)\| = \sup_{\lambda\in\sigma(A)} |\phi(\lambda)| \;.$$

(b) $\phi(A)^* = \bar{\phi}(A)$, where $\bar{\phi}(\lambda) = \overline{\phi(\lambda)}$.

(c) If $\phi(\lambda) = \phi_1(\lambda)\phi_2(\lambda)$, then $\phi(A) = \phi_1(A)\phi_2(A)$.

(d) If $\phi(\lambda) = \alpha_1\phi_1(\lambda) + \alpha_2\phi_2(\lambda)$, then $\phi(A) = \alpha_1\phi_1(A) + \alpha_2\phi_2(A)$.

(e) $\phi(A)$ is normal, i.e. $\phi(A)\phi(A)^* = \phi(A)^*\phi(A)$.

(f) If $B \in \mathcal{B}(\mathcal{H})$ commutes with A, then $\phi(A)B = B\phi(A)$.

Proof : The proof is left as exercise.

PROPOSITION 4.6 : (*Unitary one-parameter group*) Let $\phi_t(\lambda) = \exp(-i\lambda t)$. Then $\{\phi_t(A)\}$ forms a strongly continuous unitary one-parameter group whose infinitesimal generator is A.

Proof :

(i) *Group* :

$$\phi_t(A)\phi_\tau(A) = e^{-iAt}\, e^{-iA\tau} = e^{-iA(t+\tau)} = \phi_{t+\tau}(A)$$

by Proposition 4.5 (c). In particular

$$[\phi_t(A)]^{-1} = \phi_{-t}(A) .$$

(ii) *Unitarity* : This part is established by Proposition 4.5(b) and (i) : we have

$$\phi_t(A)^* \phi_t(A) = e^{iAt}\, e^{-iAt} = \phi_o(A) = I \quad .$$

Now by using Proposition 4.5 (e) we get

$$\phi_t(A)\phi_t(A)^* = \phi_t(A)^* \phi_t(A) = I \quad .$$

Thus $\phi_t(A)$ is unitary.

(iii) *Strong continuity* :

$$(\phi_{t+\tau}(A)f,g) - (\phi_t(A)f,g) = \int_{-\infty}^{\infty} [e^{i\lambda(t+\tau)} - e^{i\lambda t}]\, d(f,E_\lambda g) .$$

Since $|e^{i\lambda(t+\tau)} - e^{i\lambda t}| \leq 2$ and converges to zero as $\tau \to 0$, the integral converges to zero as $\tau \to 0$ by the Lebesgue dominated convergence theorem. Thus $\{\phi_t(A)\}$ is weakly continuous. Now

$$\|\phi_{t+\tau}(A)f - \phi_t(A)f\|^2 = \|\phi_{t+\tau}(A)f\|^2$$

$$- (\phi_{t+\tau}(A)f,\phi_t(A)f) - (\phi_t(A)f,\phi_{t+\tau}(A)f) + \|\phi_t(A)f\|^2 .$$

Since $\|\phi_s(A)f\|^2 = \|f\|^2$ and $\phi_t(A)$ is weakly continuous, the above converges to zero as $\tau \to 0$. Hence $\{\phi_t(A)\}$ is strongly continuous.

(iv) *Infinitesimal generator* :

$$\left.\begin{aligned} &\underset{\tau\to 0}{\text{s-lim}}\ \frac{1}{\tau}\,[\phi_\tau(A) - \phi_0(A)\, f] \text{ exists if and only if} \\ &\qquad f \in D(A) \text{ and equals } - iAf\ . \end{aligned}\right\} \quad (4.4)$$

Let $f \in D(A)$. Then

$$\left\|\frac{1}{\tau}\,[\phi_\tau(A)-I]f+iAf\right\|^2 = \int_{-\infty}^{\infty} [\frac{1}{\tau}\,(e^{-i\tau\lambda}-1)+i\lambda]\ d(f,E_\lambda f)\ .$$

Now $$\left|\frac{1}{\tau}\,(e^{-i\tau\lambda}-1)+i\lambda\right| = \left|\int_0^\lambda e^{-i\tau\mu}-1\ d\mu\right| \le \left|2\int_0^\lambda d\mu\right| = 2|\lambda|.$$

Since $\int_{-\infty}^{\infty} \lambda^2\ d(f,E_\lambda f) < \infty$, the above integral converges to zero by the Lebesgue dominated convergence theorem. Thus $i\,\frac{d}{d\tau}\,\phi_\tau(A)f\Big|_{\tau=0} = Af$ for $f \in D(A)$.

By Stone's theorem, if $\{U_t\}$ is any strongly continuous one-parameter unitary group, the set $\{f|\,i\,\frac{d}{d\tau}\,U_\tau f\Big|_{\tau=0}$ exists strongly$\}$ is exactly the domain of a self-adjoint operator. Since A above is self-adjoint, the statement (4.4) follows. ∎

We now come to the subdivisions of the spectrum of a self-adjoint operator A. We first show :

PROPOSITION 4.7 : Let $A = A^*$. Then (a) all eigenvalues of A are real, (b) if $Af_1 = \lambda_1 f_1$ and $Af_2 = \lambda_2 f_2$ with $\lambda_1 \neq \lambda_2$, then f_1 is orthogonal to f_2.

Proof :

(a) Suppose $Af = \lambda f$, $f \neq 0$. Then $\lambda(f,f) = (f,Af) = (Af,f) = (\lambda f,f) = \bar{\lambda}(f,f)$, i.e. $\lambda = \bar{\lambda}$ since $(f,f) \neq 0$.

(b) As above

$$(\lambda_1 - \lambda_2)(f_1,f_2) = (Af_1,f_2) - (f_1,Af_2) = 0 .$$

Since $\lambda_1 \neq \lambda_2$, $(f_1,f_2) = 0$. ■

DEFINITION : The *point spectrum* $\sigma_p(A)$ of A is the set of all eigenvalues of A. $\mathcal{H}_p(A)$ is the (closed) subspace spanned by all eigenvectors of A, $\mathcal{H}_c(A)$ the orthogonal complement of $\mathcal{H}_p(A)$.

By Proposition (4.7)(b), *the point spectrum is a countable set,* since the separable space $\mathcal{H}$ has at most a countable family of mutually orthogonal vectors. By Lemma 4.3, $\lambda \in \sigma_p(A) \Leftrightarrow E_{\{\lambda\}} \neq 0$. The projection $E_p(A)$ with range $\mathcal{H}_p(A)$ may be written as

$$E_p(A) = \sum_i E_{\{\lambda_i\}} , \quad \lambda_i \in \sigma_p(A)$$

and since each $E_{\{\lambda_i\}}$ commutes with A, $E_p(A)$ also commutes with A. We define

$$A_p \equiv AE_p(A) = E_p(A)\, AE_p(A) .$$

Let $E_c(A) \equiv I - E_p(A)$, the projection with range $\mathcal{H}_c(A)$. $E_c(A)$ also commutes with A, and we set

$$A_c \equiv AE_c(A) = E_c(A)\, AE_c(A) .$$

A_c, viewed as an operator in $\mathcal{H}_c(A)$, is called the *continuous part of* A, and its spectrum is called the *continuous spectrum* $\sigma_c(A)$ of A :

$$\sigma_c(A) \equiv \sigma(A_c) .$$

$\sigma_c(A)$ is a closed subset of $\mathbb{R}$, since it is the spectrum of the self-adjoint operator A_c. $\sigma_p(A)$ on the other hand need not be closed, since it is not necessarily the spectrum of an operator. In fact, $\overline{\sigma_p(A)} = \sigma(A_p)$. Also $\sigma(A) = \overline{\sigma_p(A)} \cup \sigma_c(A)$. $\sigma_p(A)$ and $\sigma_c(A)$ need not be disjoint sets, however.

If $\mathcal{H}_p(A) = \mathcal{H}$, A is said to have *pure point spectrum*. If $\mathcal{H}_c(A) = \mathcal{H}$, A has *purely continuous spectrum*.

We define a further subdivision of $\mathcal{H}$. If $f \in \mathcal{H}$, then $m_f(\Delta) = (f, E_\Delta f) = \|E_\Delta f\|^2$ is a non-negative countably additive Borel measure on $\mathbb{R}$. By the Lebesgue decomposition Theorem, m_f has a unique decomposition into $m_f = m_{f,ac} + m_{f,s}$, where $m_{f,ac}$ is absolutely continuous and $m_{f,s}$ singular with respect to the Lebesgue measure $|.|$.

DEFINITION : f is *absolutely continuous* (resp. *singular*) *with respect to* A if $m_f = m_{f,ac}$ (resp. $m_f = m_{f,s}$). The set of all f in $\mathcal{H}$ which are absolutely continuous (resp. singular) with respect to A is denoted $\mathcal{H}_{ac}(A)$ [resp. $\mathcal{H}_s(A)$] and called the *subspace of absolute continuity* (resp. of *singularity*) of A, [AS], [K].

PROPOSITION 4.8 : $\mathcal{H}_{ac}(A)$ and $\mathcal{H}_s(A)$ are (closed) subspaces and orthogonal complements to each other (i.e. $\mathcal{H} = \mathcal{H}_{ac}(A) \oplus \mathcal{H}_s(A)$). A commutes with $E_{ac}(A)$ and $E_s(A)$, the projections with range $\mathcal{H}_{ac}(A)$ and $\mathcal{H}_s(A)$ respectively.

Before proving this result, we give another definition and some examples.

DEFINITION : The *absolutely continuous spectrum* $\sigma_{ac}(A)$ (resp. the *singular spectrum* $\sigma_s(A)$) is the spectrum of the operator $AE_{ac}(A)$ viewed as an operator in $\mathcal{H}_{ac}(A)$ (resp. of $AE_s(A)$ viewed as an operator in $\mathcal{H}_s(A)$).

Both $\sigma_{ac}(A)$ and $\sigma_s(A)$ are closed subsets of $\mathbb{R}$. They need not be disjoint.

If $Af = \lambda f$, then $E_{\{\lambda\}} f = f$ by Lemma 4.3, hence $m_f(\{\lambda\}) = \|f\|^2 \neq 0$ and $m_f(\Delta) = 0$ if $\lambda \notin \Delta$. Thus $f \in \mathcal{H}_s(A)$, $\lambda \in \sigma_s(A)$. Thus

$$\mathcal{H}_p(A) \subseteq \mathcal{H}_s(A) \quad , \quad \sigma_p(A) \subseteq \sigma_s(A) \quad .$$

By taking the complements in the above inclusion :

$$\mathcal{H}_s(A)^{\perp} = \mathcal{H}_{ac}(A) \subseteq \mathcal{H}_p(A)^{\perp} = \mathcal{H}_c(A) \ .$$

Thus $\sigma_{ac}(A) \subseteq \sigma_c(A)$.

We define $\mathcal{H}_{sc}(A)$ as the complement of $\mathcal{H}_{ac}(A)$ in $\mathcal{H}_c(A)$, i.e.

$$\mathcal{H} = \mathcal{H}_p(A) \oplus \mathcal{H}_{ac}(A) \oplus \mathcal{H}_{sc}(A) \ . \tag{4.5}$$

Thus $\mathcal{H}_{sc}(A) = \mathcal{H}_s(A) \cap \mathcal{H}_c(A)$. The spectrum of $AE_{sc}(A)$ (viewed as an operator in $\mathcal{H}_{sc}(A)$), is called the *singularly continuous spectrum* $\sigma_{sc}(A)$ of A.

DEFINITION : The set of all isolated (in $\sigma(A)$) eigenvalues of finite multiplicity is called the *discrete spectrum* $\sigma_d(A)$ of A. The set $\sigma(A)\backslash\sigma_d(A)$ is called the *essential spectrum* $\sigma_e(A)$.

In other words :

$\lambda \in \sigma_d(A) \Leftrightarrow \lambda$ is isolated in $\sigma(A)$ and $0 < \dim E_{\{\lambda\}} < \infty$.

$\lambda \in \sigma_e(A) \Leftrightarrow \dim_{(\lambda-\varepsilon,\lambda+\varepsilon]}\mathcal{H} = \infty$ for each $\varepsilon > 0$.

The essential spectrum consists of the continuous spectrum, the eigenvalues of infinite multiplicity and the accumulation points of the point spectrum.

Examples :

4.1. *Operators with pure point spectrum.*

Let dim $\mathcal{H} = \infty$, let $\{e_k\}_{k=1}^{\infty}$ be an orthonormal basis of $\mathcal{H}$.

(a) Set $A_o(\sum_{k=1}^{N} \alpha_k e_k) = \sum_{k=1}^{N} k\, \alpha_k e_k$, for all $N < \infty$, $\alpha_k \in \mathbb{R}$.

A_o is symmetric. Let $A = \bar{A}_o$. A is self-adjoint but unbounded.

$$\sigma(A) = \sigma_p(A) = \sigma_d(A) = \mathbb{Z}^+ .$$

$$\sigma_c(A) = \sigma_e(A) = \phi .$$

(b) Define A by $Ae_k = \frac{1}{k} e_k$ and extend it by linearity and continuity to $\mathcal{H}$. $A \in \mathcal{B}(\mathcal{H})$, $\|A\| = 1$.

$$\sigma_p(A) = \sigma_d(A) = \{\tfrac{1}{k} \mid k = 1,2,\ldots\}.$$

$$\sigma_c(A) = \phi .$$

$$\sigma(A) = \overline{\sigma_p(A)} = \{\tfrac{1}{k}, 0\} = \sigma_p(A) \cup \{0\} .$$

$$\sigma_e(A) = \{0\}.$$

(c) Let $\{r_j\}_{j=1}^{\infty}$ be an enumeration of all rationals in (0,1). Define $Ae_k = r_k e_k$ and extend by linearity and continuity to $\mathcal{H}$.

$\sigma_p(A)$ is the set of all rationals in (0,1) .

$\sigma_d(A) = \phi$ (no eigenvalue is isolated) .

$\sigma_c(A) = \phi$.

$\sigma(A) = \overline{\sigma_p(A)} = \sigma_e(A) = [0,1]$.

(d) If A is a compact self-adjoint operator, it has only point spectrum, say $\sigma_p(A) = \{\lambda_i\}$. Each $\lambda_i \neq 0$ is of finite multiplicity, i.e. $\sigma_d(A) = \sigma_p(A)$ if $0 \notin \sigma_p(A)$ and $\sigma_d(A) = \sigma_p(A)\setminus\{0\}$ if $0 \in \sigma_p(A)$. The point $\lambda = 0$ belongs to $\sigma_e(A)$: $\sigma_e(A) = \{0\}$. $\lambda = 0$ may be an accumulation point of $\sigma_p(A)$ or an eigenvalue of infinite multiplicity or both, [AS], [RS, vol. I].

4.2. *Purely continuous operators.*

(a) Let $\mathcal{H} = L^2([0,1],d\mu)$ and define A by $(Af)(x) = xf(x)$. Then $\|A\| \leq 1$. Spectral family of A :

$$0 \leq \lambda \leq 1 : (E_\lambda f)(x) = \begin{cases} f(x) & 0 \leq x \leq \lambda \\ 0 & x > \lambda. \end{cases}$$

$$\lambda < 0 : E_\lambda = 0$$

$$\lambda \geq 1 : E_\lambda = I .$$

Indeed :

$$\int_{\mathbb{R}} \lambda \, d(f,E_\lambda g) = \int_0^1 \lambda \, d \int_{0\leq x\leq\lambda} \overline{f(x)} g(x) d\mu(x)$$

$$= \int_0^1 \lambda \, \overline{f(\lambda)} g(\lambda) d\mu(\lambda) = (Af,g) .$$

The spectral properties of A depend on the measure μ.

(α) Let μ be the Lebesgue measure :

$$\sigma(A) = \sigma_c(A) = \sigma_{ac}(A) = \sigma_e(A) = [0,1] .$$

$$\sigma_p(A) = \sigma_s(A) = \phi.$$

(β) If μ is a singularly continuous monotone function (i.e. μ continuous, $\mu(b) > \mu(a)$ if $b > a$, $\mu'(x) = 0$ a.e. with respect to Lebesgue measure [RN],) then

$$\sigma_c(A) = \sigma_{sc}(A) = \sigma_e(A) = [0,1] ,$$

$$\sigma_p(A) = \sigma_{ac}(A) = \phi .$$

(b) Let $\mathcal{H} = L^2(\mathbb{R}^n)$, $A = K_o$ where K_o is the operator that we introduced in Example 3.3. The Fourier transformation diagonalizes K_o :

$$(\widetilde{K_o f})(k) = \sum_{j=1}^{n} k_j^2 \tilde{f}(k) \equiv |k|^2 \tilde{f}(k) .$$

As above :

$$0 \leq \lambda : (\widetilde{E_\lambda^o f})(k) = \begin{cases} \tilde{f}(k) & |k|^2 \leq \lambda \\ 0 & |k|^2 > \lambda. \end{cases}$$

$$\lambda < 0 : E_\lambda^o = 0 .$$

In spherical polar coordinates in $\mathbb{R}^n$, with $\lambda = k^2$, $d\omega$ = Lebesgue measure on S^{n-1} :

$$dk = k^{n-1} \, dk \, d\omega = \frac{1}{2} \lambda^{\frac{1}{2}(n-2)} \, d\lambda \, d\omega .$$

We have $(f, E_\mu^o g) = \int_{|k|^2 \leq \mu} \overline{\tilde{f}(k)} \tilde{g}(k) \, dk = \frac{1}{2} \int_0^\mu \lambda^{\frac{n-2}{2}} \, d\lambda \int_{S^{n-1}} d\omega \, \overline{\tilde{f}(\sqrt{\lambda}\omega)} \tilde{g}(\sqrt{\lambda}\omega)$.

Therefore each $f \in \mathcal{H}$ is in $\mathcal{H}_{ac}(K_o)$, since $\|E_\lambda^o f\|^2$ is an absolutely continuous function. Thus K_o *is absolutely continuous*,

$$\boxed{\sigma(K_o) = \sigma_{ac}(K_o) = \sigma_e(K_o) = [0,\infty)} \quad .$$

Spectral representation of K_o.

The operator $\mathcal{U}_o : L^2(\mathbb{R}^n) \to L^2([0,\infty), L^2(S^{n-1}))$ is defined as [AS] :

$$(\mathcal{U}_o f)_\lambda(\omega) = 2^{-\frac{1}{2}} \lambda^{\frac{n-2}{4}} \tilde{f}(\sqrt{\lambda}\omega) \qquad (\omega \in S^{n-1}) \tag{4.6}$$

[i.e. for a.e. $\lambda \in [0,\infty)$, $(\mathcal{U}_o f)_\lambda$ is a vector in $L^2(S^{n-1})$]. $\mathcal{U}_o$ is unitary and called a *spectral transformation* for K_o, since

$$(\mathcal{U}_o K_o f)_\lambda = \lambda (K_o f)_\lambda \ .$$

<u>Proof of Proposition 4.8</u> : [K]

(i) Let $f \in \mathcal{H}_{ac}$, $g \in \mathcal{H}_s$. There is a Borel set Δ_o with $|\Delta_o| = 0$ such that $E_{\Delta_o} g = g$. Thus

$$(f,g) = (f,E_{\Delta_o} g) = (E_{\Delta_o} f,g) = (0,g) = 0 \ ,$$

since $f \in \mathcal{H}_{ac}$. Therefore $\mathcal{H}_{ac} \perp \mathcal{H}_s$.

(ii) Let $f \in \mathcal{H}$ and decompose $m_f = m_{f,ac} + m_{f,s}$. There is a Borel set Δ_o with $|\Delta_o| = 0$ such that for each Δ :

$$m_{f,s}(\Delta) = m_{f,s}(\Delta \cap \Delta_o) \ .$$

Set $g = E_{\Delta_o} f$, $h = f - g$. Then by (4.1)

$$m_g(\Delta) = \|E_\Delta g\|^2 = \|E_\Delta E_{\Delta_0} f\|^2 = \|E_{\Delta\cap\Delta_0} f\|^2$$

$$= m_f(\Delta\cap\Delta_0) = m_{f,s}(\Delta)$$

since $m_{f,ac}(\Delta\cap\Delta_0) = 0$. This implies $g \in \mathcal{H}_s$.

$$m_h(\Delta) = \|E_\Delta h\|^2 = \|E_\Delta [I-E_{\Delta_0}]f\|^2$$

$$= \|E_\Delta f\|^2 - 2\|E_{\Delta\cap\Delta_0} f\|^2 + \|E_{\Delta\cap\Delta_0} f\|^2$$

$$= m_f(\Delta) - m_f(\Delta\cap\Delta_0) = m_f(\Delta) - m_{f,s}(\Delta) = m_{f,ac}(\Delta) .$$

This implies that $h \in \mathcal{H}_{ac}$.

Thus each $f \in \mathcal{H}$ may be written as $f = g + h$, $g \in \mathcal{H}_s$, $h \in \mathcal{H}_{ac}$, and $\mathcal{H}_s \perp \mathcal{H}_{ac}$. Thus $\mathcal{H} = \mathcal{H}_s \oplus \mathcal{H}_{ac}$.

(iii) Let $f \in \mathcal{H}_{ac}$. Then $E_\Delta E_\lambda f = E_\lambda E_\Delta f = 0$ if $|\Delta| = 0$, i.e. $E_\lambda f \in \mathcal{H}_{ac}$. Thus $E_\lambda E_{ac}(A) = E_{ac}(A)E_\lambda$, hence also

$$E_\lambda E_s(A) = E_s(A)E_\lambda .$$

(iv) Let $f \in D(A)$, and write $f = g + h$, $g \in \mathcal{H}_s$, $h \in \mathcal{H}_{ac}$. Then

$$\|Af\|^2 = \int \lambda^2 \, d(f,E_\lambda f) = \int \lambda^2 \, d(g,E_\lambda g) + \int \lambda^2 \, d(h,E_\lambda h) ,$$

since $(g,E_\lambda h) = 0$ by (iii).
Thus $g \in D(A)$ and $h \in D(A)$.

Let $\phi \in \mathcal{H}$. Then by (iii) above, one has

$$(AE_{ac}f,\phi) = \int \lambda \, d(h,E_\lambda\phi)$$

$$(E_{ac}Af,\phi) = (Af,E_{ac}\phi) = \int \lambda \, d(f,E_\lambda E_{ac}\phi)$$

$$= \int \lambda \, d(E_{ac}f,E_\lambda\phi) = \int \lambda \, d(h,E_\lambda\phi) \; .$$

Hence $E_{ac}A \subseteq AE_{ac}$, i.e. A commutes with E_{ac}. Similarly $E_sA \subseteq AE_s$. ∎

The next lemma justifies the designation "*continuous spectrum*".

LEMMA 4.9 :

(a) $f \in \mathcal{H}_c(A)$ if and only if $\lambda \mapsto (f,E_\lambda f)$ is continuous.

(b) Let Δ be a countable subset of $\mathbb{R}$. Then $E_\Delta\mathcal{H}_c(A) = \{0\}$, i.e. the range of E_Δ is contained in $\mathcal{H}_p(A)$.

Proof :

(a) If $(f,E_\lambda f)$ is continuous, $(f,E_{\{\lambda\}}f) = 0$ for all $\lambda \in \mathbb{R}$, i.e. $\| E_{\{\lambda\}}f\|^2 = 0$.

Let g be an eigenvector of A, i.e. $g = E_{\{\lambda\}}g$ for some λ. Then $(f,E_{\{\lambda\}}g) = (E_{\{\lambda\}}f,g) = (0,g) = 0$.
Therefore $f \perp \mathcal{H}_p(A)$, i.e. $f \in \mathcal{H}_c(A)$.

If $f \in \mathcal{H}_c(A)$, then $f \perp E_{\{\lambda\}}f$ for all λ, i.e. $(f,E_{\{\lambda\}}f) = 0$ or $\lim_{\varepsilon \to +0} (f,E_{\lambda-\varepsilon}f) = (f,E_\lambda f)$. Since $\{E_\lambda\}$ is also right-continuous, $\lambda \mapsto (f,E_\lambda f)$ is continuous.

(b) We have $\Delta = \bigcup_{i=1}^{\infty} \{\lambda_i\}$, thus $E_\Delta f = \sum_i E_{\{\lambda_i\}}f$ for each $f \in \mathcal{H}$. Now $E_{\{\lambda_i\}}f$ is zero or an eigenvector of A [see Lemma 4.3], so that $E_\Delta f \in \mathcal{H}_p(A)$. ∎

We now prove some simple spectral properties of Schrödinger operators.

DEFINITION : A self-adjoint operator A is *bounded below* if there is $M > -\infty$ such that $z \in \rho(A)$ for all $z < M$. The least upper bound of all such M is called the *lower bound* $M(A)$ of A. ($M(A)$ coincides with the lower bound of $\sigma(A)$, Problem 4.7).

PROPOSITION 4.10 : Let A be self-adjoint and bounded below, B symmetric and A-bounded with A-bound $\beta_0 < 1$. Then the self-adjoint operator $A + B$ is bounded below.

Proof : One has $A + B - z = [I + B(A-z)^{-1}](A-z)$, i.e.

$$[A+B-z]^{-1} = (A-z)^{-1} [I+B(A-z)^{-1}]^{-1} .$$

$(A-z)^{-1}$ is in $\mathcal{B}(\mathcal{H})$ if $z < M(A)$. $[(I+B(A-z)^{-1}]^{-1} \in \mathcal{B}(\mathcal{H})$ if $\|B(A-z)^{-1}\| < 1$ [Proposition 2.3]. Thus it suffices to show that there exists $M \le M(A)$ such that $\|B(A-z)^{-1}\| < 1$ for all $z < M$.

Let $g \in \mathcal{H}$, $f = (A-z)^{-1} g \in D(A)$. Then

$$\|B(A-z)^{-1}g\| = \|Bf\| \le \alpha\|(A-z)^{-1}g\| + \beta\|A(A-z)^{-1}g\| .$$

We obtain $$\|B(A-z)^{-1}\| \le \alpha\|(A-z)^{-1}\| + \beta\|A(A-z)^{-1}\| .$$

Now by Proposition 4.5(a) it follows that
$\|(A-z)^{-1}\| = \sup_{\lambda \ge M(A)} |\lambda-z|^{-1} = (M(A) - z)^{-1}$ and

$$\|A(A-z)^{-1}\| = \sup_{\lambda \ge M(A)} \frac{|\lambda|}{|\lambda-z|} = \max\{1, \frac{|M(A)|}{M(A)-z}\} .$$

Since $\beta < 1$, we have $\|B(A-z)^{-1}\| < 1$ for all $z < z_0$, where

$z_0 = M(A) - \max \{\frac{\alpha}{1-\beta}, \alpha+\beta\ |M(A)|\}$. ∎

<u>LEMMA 4.11</u> : Let $A = A^*$, $\mu \in \mathbb{R}$. Then $\mu \in \sigma_e(A)$ if and only if there exists a sequence $\{f_n\} \in D(A)$ such that $\|f_n\| = 1$, $\text{w-}\lim_{n\to\infty} f_n = 0$ and $\text{s-}\lim_{n\to\infty} (A-\mu)f_n = 0$.

<u>Proof</u> :

(i) "*If*" : Let $a < \mu < b$. Then

$$\|(A-\mu)f_n\|^2 = \int_{-\infty}^{\infty} (\lambda-\mu)^2\, d\,\|E_\lambda f_n\|^2 \geq (a-\mu)^2 \int_{-\infty}^{a} d\,\|E_\lambda f_n\|^2$$

$$+ (b-\mu)^2 \int_{b}^{\infty} d\,\|E_\lambda f_n\|^2 = (a-\mu)^2\, \|E_a f_n\|^2 + (b-\mu)^2\, \|(I-E_b)f_n\|^2 .$$

Since $\|(A-\mu)f_n\| \to 0$, we have $\|E_a f_n\| \to 0$ and $\|(I-E_b)f_n\| \to 0$. Now by using the fact that $|\|f\| - \|g\|| \leq \|f-g\|$ one has

$$|\|f_n\| - \|E_{(a,b]} f_n\|| \leq \|f_n - E_{(a,b]} f_n\| = \|(I-E_b)f_n + E_a f_n\|$$

$$\leq \|(I-E_b)f_n\| + \|E_a f_n\| \to 0 \quad \text{as } n \to \infty,$$

i.e. $\lim_{n\to\infty} \|E_{(a,b]} f_n\| = 1 = \|f_n\|$.

Assume that $\dim E_{(a,b]}\mathcal{H} = N < \infty$, and let $e_1,\ldots,e_N$ be an orthonormal basis of $E_{(a,b]}\mathcal{H}$. Then

$$\|E_{(a,b]} f_n\|^2 = \sum_{k=1}^{N} |(e_k,f_n)|^2 .$$

Since $N < \infty$ and $(e_k,f_n) \to 0$ for each k [because $\underset{n\to\infty}{\text{w-lim}}\, f_n = 0$], we have

$$\lim_{n\to\infty} \|E_{(a,b]} f_n\| = 0 ,$$

a contradiction. Hence $\dim E_{(a,b]}\mathcal{H} = \infty$, i.e. $\mu \in \sigma_e(A)$.

(ii) "*Only if*" : If $\mu \in \sigma_e(A)$, then $\dim E_{(\mu-\frac{1}{k},\mu+\frac{1}{k}]}\mathcal{H} = \infty$ for each $k = 1,2\ldots$. Thus there exists an infinite orthonormal sequence $\{f_n\}$ such that $f_k \in E_{(\mu-\frac{1}{k},\mu+\frac{1}{k}]}\mathcal{H}$. Then $\|f_n\| = 1$ and $\underset{n\to\infty}{\text{w-lim}}\, f_n = 0$ [by Example 1 in chapter 2]. Also

$$\|(A-\mu)f_n\|^2 = \|(A-\mu)\, E_{(\mu-\frac{1}{n},\mu+\frac{1}{n}]} f_n\|^2$$

$$\int_{\mu-\frac{1}{n}}^{\mu+\frac{1}{n}} (\lambda-\mu)^2 \, d\,\|E_\lambda f_n\| \le \frac{1}{n^2}\int d\,\|E_\lambda f_n\|^2 = \frac{1}{n^2}\,\|f_n\|^2 = \frac{1}{n^2} ,$$

i.e. $\underset{n\to\infty}{\text{s-lim}}\, (A-\mu)f_n = 0$. ∎

First resolvent equation : Let $A = A^*$, $z_1,z_2 \in \rho(A)$. Then

$$(A-z_1)^{-1} - (A-z_2)^{-1} = (z_1-z_2)(A-z_1)^{-1}\,(A-z_2)^{-1} . \qquad (4.7)$$

Proof : Let $R_z \equiv (A-z)^{-1}$. R_z maps $\mathcal{H}$ onto $D(A)$. Hence $AR_z f$ is defined for each $f \in \mathcal{H}$, and

$$AR_z = (A-z)R_z + zR_z = I + zR_z \in \mathcal{B}(\mathcal{H}) .$$

Now

$$R_{z_1} - R_{z_2} = R_{z_1} (A-z_2)R_{z_2} - R_{z_1}(A-z_1)R_{z_2} = (z_1-z_2)R_{z_1} R_{z_2} . \blacksquare$$

Second resolvent equation : Let $H_0^* = H_0$, $H = H^*$ and $D(H_0) = D(H)$. Set $V = H - H_0$. Then, if $z \in \rho(H_0) \cap \rho(H)$:

$$(H-z)^{-1} - (H_0-z)^{-1} = -(H-z)^{-1} V(H_0-z)^{-1} = -(H_0-z)^{-1} V(H-z)^{-1}. \quad (4.8)$$

Proof : Notice that $V(H_0-z)^{-1}f$, $H(H_0-z)^{-1}f$ and $H_0(H_0-z)^{-1}f$ are defined for each $f \in \mathcal{H}$. Hence

$$\begin{aligned}(H-z)^{-1}f - (H_0-z)^{-1}f \\ &= (H-z)^{-1} (H_0-z)(H_0-z)^{-1}f - (H-z)^{-1}(H-z)(H_0-z)^{-1}f \\ &= - (H-z)^{-1} V(H_0-z)^{-1}f ,\end{aligned}$$

and similarly for the other relation. $\blacksquare$

LEMMA 4.12 : Let $A = A^*$ and $D(B) \supseteq D(A)$. If $B(A-z_0)^{-1} \in \mathcal{B}(\mathcal{H})$ [resp. $B(A-z_0)^{-1} \in \mathcal{B}_\infty$] for some $z_0 \in \rho(A)$, then $B(A-z)^{-1} \in \mathcal{B}(\mathcal{H})$ [resp. $B(A-z)^{-1} \in \mathcal{B}_\infty$] for each $z \in \rho(A)$.

Proof : By the first resolvent equation

$$BR_z = BR_{z_0} + (z-z_0)BR_{z_0} R_z .$$

From $BR_{z_0} \in \mathcal{B}(\mathcal{H})$ [resp. $\in \mathcal{B}_\infty$] and Lemma 2.4, one obtains that $BR_z \in \mathcal{B}(\mathcal{H})$ [resp. $\in \mathcal{B}_\infty$]. $\blacksquare$

LEMMA 4.13 : Let $H_0 = H_0^*$, let V be H_0-bounded with H_0-bound $\beta_0 < 1$ and let $H = H_0 + V$. Then

(a) $V(H_0-z)^{-1} \in \mathcal{B}(\mathcal{H})$ and $V(H-z)^{-1} \in \mathcal{B}(\mathcal{H})$

for each $z \in \rho(H_0) \cap \rho(H)$.

(b) If V is H_0-compact, then

$V(H_0-z)^{-1} \in \mathcal{B}_\infty$ and $V(H-z)^{-1} \in \mathcal{B}_\infty$

for each $z \in \rho(H_0) \cap \rho(H)$.

Remark : The hypothesis of (b) implies that V is H_0-bounded with H_0-bound $\beta_0 = 0$ by Proposition 3.9.

Proof :

(a) We have seen in an earlier proof [Proposition 3.6] that there exists $\lambda > 0$ such that

$$\|\lambda V(\lambda H_0 \pm i)^{-1}\| = \|V(H_0 \pm \tfrac{i}{\lambda})^{-1}\| < 1 .$$

Thus $V(H_0-z)^{-1} \in \mathcal{B}(\mathcal{H})$ for all $z \in \rho(H_0)$.

Now $H_0 + V-z = [I+V(H_0-z)^{-1}](H_0-z)$. Now by using the Neumann series , Proposition 2.3, one obtains $[I+V(H_0-z)^{-1}]^{-1} \in \mathcal{B}(\mathcal{H})$ for $z = \frac{i}{\lambda}$. Thus $(H-z)^{-1} = (H_0-z)^{-1} [I+V(H_0-z)^{-1}]^{-1} \in \mathcal{B}(\mathcal{H})$ for $z = \pm \frac{i}{\lambda}$, hence for all $z \in \rho(H)$.

(b) if V is H_0-compact, then $V(H_0-z_0)^{-1} \in \mathcal{B}_\infty$ for some z_0, hence for all $z \in \rho(H_0)$ by Lemma 4.12.

From the second resolvent equation

$$V(H-z)^{-1} = V(H_0-z)^{-1} - V(H_0-z)^{-1} V(H-z)^{-1} .$$

Since $V(H_o-z)^{-1} \in \mathcal{B}_\infty$, $V(H-z)^{-1} \in \mathcal{B}(\mathcal{H})$ by (a), it follows from Lemma 2.4 that : $V(H-z)^{-1} \in \mathcal{B}_\infty$ for all $z \in \rho(H) \cap \rho(H_o)$. ∎

LEMMA 4.14 : Let $\{f_n\} \in \mathcal{H}$, $\underset{n\to\infty}{\text{w-lim}}\, f_n = 0$ and $C \in \mathcal{B}_\infty$. Then $\underset{n\to\infty}{\text{s-lim}}\, Cf_n = 0$.

Proof : $\exists\, M < \infty$ such that $\|f_n\| \le M$, by the uniform boundedness principle. Let $\varepsilon > 0$, choose a finite rank operator T with $\|T-C\| < \frac{\varepsilon}{2M}$, say $Tf = \sum_{i=1}^{N} (e_i,f)h_i$, with $\|h_i\| = 1$. Then

$$\|Cf_n\| \le \|(C-T)f_n\| + \|Tf_n\| \le \frac{\varepsilon}{2} + \sum_{i=1}^{N} |(e_i,f_n)|\, \|h_i\| .$$

Since $\underset{n\to\infty}{\text{w-lim}}\, f_n = 0$, $\exists n_o$ such that $|(e_i,f_n)| < \frac{\varepsilon}{2N}$ for all $i = 1,\ldots,N$ and all $n > n_o$. Thus $\|Cf_n\| < \varepsilon$ for all $n > n_o$. ∎

PROPOSITION 4.15 : Let $A = A^*$, B be symmetric and A-compact. Then

$$\sigma_e(A+B) = \sigma_e(A) .$$

Proof :

(i) Let $\lambda \in \sigma_e(A)$. By Lemma 4.11, there exists $\{f_n\} \in D(A) = D(A+B)$ with $\|f\|_n = 1$, $\underset{n\to\infty}{\text{w-lim}}\, f_n = 0$ and $\underset{n\to\infty}{\text{s-lim}}\, (A-\lambda)f_n = 0$.

Let $z \in \rho(A)$. Then

$$(A-z)f_n = (A-\lambda)f_n + (\lambda-z)f_n .$$

Since $(A-\lambda)f_n \to 0$ strongly as $n \to \infty$ and $(\lambda-z)f_n \to 0$ weakly as $n \to \infty$, it follows that $(A-z)f_n \to 0$ weakly as $n \to \infty$. Since $B(A-z)^{-1} \in \mathcal{B}_\infty$, one has by Lemma 4.14

$$Bf_n = B(A-z)^{-1}\,(A-z)f_n \to 0 \text{ strongly as } n \to \infty.$$

Thus $(A+B-\lambda)f_n = (A-\lambda)f_n + Bf_n \to 0$ strongly as $n \to \infty$.
Therefore $\lambda \in \sigma_e(A+B)$ by Lemma 4.11, i.e. $\sigma_e(A) \subseteq \sigma_e(A+B)$.

(ii) B is also (A+B)-compact by Lemma 4.13(b). We interchange the roles of A and A+B (with $B \to -B$) in the above, leading to

$$\sigma_e(A+B) \subseteq \sigma_e(A) \;. \quad \blacksquare$$

<u>Application</u> :

$$\mathcal{H} = L^2(\mathbb{R}^n),\;\; H = K_0 + V,\; V = V(x),\; V \text{ real} \;.$$

(i) <u>n = 3</u> : $V = V_1 + V_2$ with $V_1(.) \in L^2(\mathbb{R}^3)$, $V_2(.)$ essentially bounded and $\lim_{|x|\to\infty} |V_2(x)| = 0$.
$(K_0 + i)^{-1}$ is the convolution operator by the inverse Fourier transform of $(|k|^2 + i)^{-1}$, an L^2-function. Hence $V_1(K_0 + i)^{-1}$ is Hilbert-Schmidt by Proposition 2.7. Set

$$V_{2,R}(x) = \begin{cases} V_2(x) & |x| \le R \\ 0 & |x| > R \;. \end{cases}$$

Again $V_{2,R}(K_0 + i)^{-1} \in \mathcal{B}_2$, and

$$\| V_2(K_0 + i)^{-1} - V_{2,R}(K_0 + i)^{-1} \| \le \sup_{|x|\ge R} |V_2(x)| \; \|(K_0+i)^{-1}\|$$

which converges to zero as $R \to \infty$.

Thus $V_2(K_0 + i)^{-1}$ is compact by Lemma 2.4, and V is K_0-compact.

(ii) $\underline{n > 3}$: $V = V_1 + V_2$, V_2 as above, $V_1 \in L^p(\mathbb{R}^n)$ for some $p > \frac{n}{2}$ [Example before Proposition 3.9].

K_0 is bounded below with lower bound $M(K_0) = 0$, and $\sigma_e(K_0) = \sigma(K_0) = [0,\infty)$. Hence $H = K_0 + V$ is bounded below in the above cases, $\sigma_e(H) = [0,\infty)$. The spectrum of H in $[M(H),0)$ consists only of isolated eigenvalues of finite multiplicity, which possibly accumulate at $\lambda = 0$:

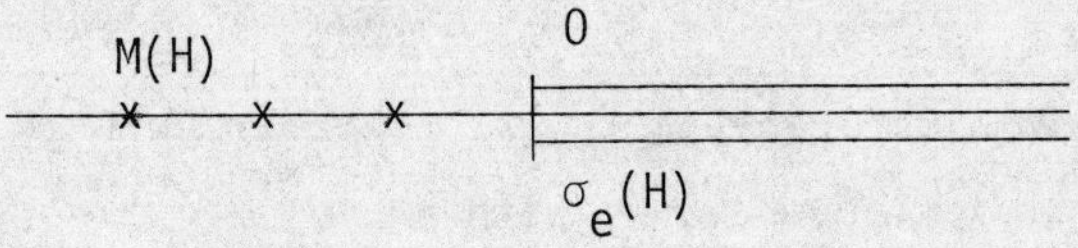

or

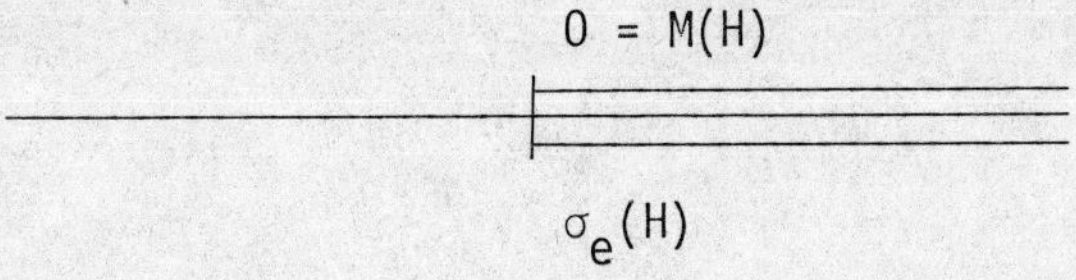

or

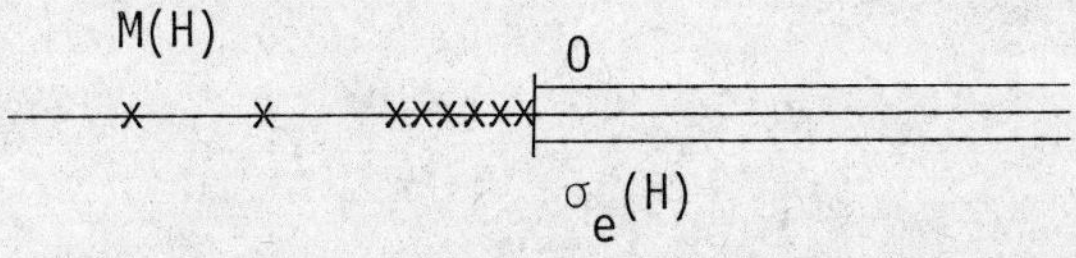

The study of the nature of $\sigma_e(H)$ is more difficult and will be done in the following chapters. Indeed, even if $\sigma_e(A)$ is invariant under a compact or A-compact perturbation B, its nature may completely change even if B is very small. We cite the following result :

Weyl-von Neumann Theorem : Let $A = A^*$ and $\varepsilon > 0$. There exists a self-adjoint Hilbert-Schmidt operator B with $\|B\|_{HS} < \varepsilon$ such that A + B has pure point spectrum, [K].

Thus, if e.g. $\sigma_e(A) = \sigma_{ac}(A) = [0,1]$, A + B may have only eigenvalues. Since $\sigma_e(A+B) = \sigma_e(A)$, these eigenvalues must be everywhere dense in [0,1].

5 Integration in Hilbert space

We need two types of integrals : integrals with respect to Lebesgue measure and integrals with respect to a spectral measure. It is enough to do Riemann-type integrals, since we shall have continuous integrands. We shall not prove results about the first type of integral, since these proofs are standard in integration theory.

Let $[a,b]$ be a finite interval in $\mathbb{R}$. A *partition* Π of $[a,b]$ is a set of numbers $\{s_0,s_1,\ldots,s_n ; u_1,\ldots,u_n\}$ with $a = s_0 < s_1 <\ldots< s_n = b$ and $u_k \in (s_{k-1},s_k]$. We define

$$|\Pi| = \max_{k=1,\ldots,n} |s_k-s_{k-1}|.$$

Let $f : [a,b] \to \mathcal{H}$ be a vector-valued function. Let

$$\sum_\Pi(f) = \sum_{k=1}^{n} (s_k-s_{k-1})f(u_k) \quad . \tag{5.1}$$

Let $\{\Pi_r\}_{r=1}^{\infty}$ be a sequence of partitions with $\lim_{r\to\infty} |\Pi_r| = 0$. One defines

$$\int_a^b f(s)\, ds = \operatorname*{s-lim}_{r\to\infty} \sum\nolimits_{\Pi_r}(f) \tag{5.2}$$

if the limit exists and is independent of the sequence $\{\Pi_r\}$. We write simply

$$\int_a^b f(s)\, ds = \operatorname*{s-lim}_{|\Pi|\to 0} \sum\nolimits_\Pi(f) \ . \tag{5.3}$$

Improper integrals are defined as strong limits of proper ones. For example

$$\int_a^\infty f(s)\, ds = \operatorname{s-lim}_{b\to\infty} \int_a^b f(s)\, ds \quad . \tag{5.4}$$

<u>PROPOSITION 5.1</u> : Let a,b be finite or infinite.

(a) $\|\int_a^b f(s)\, ds\| \le \int_a^b \|f(s)\|\, ds$. (from the triangle inequality) (5.5)

(b) If $A \in \mathcal{B}(\mathcal{H})$, then $A \int_a^b f(s)\, ds = \int_a^b Af(s)\, ds$. (5.6)

(c) If f is strongly continuous, then $\int_c^d f(s)\, ds$ exists for all (5.7)
finite c,d. If c or d is infinite, then $\int_c^d f(s)\, ds$ exists provided
that $\int_c^d \|f(s)\|\, ds < \infty$. (5.8)

(d) If f is strongly differentiable with strongly continuous and integrable derivative f', then

$$\int_a^b f'(s)\, ds = f(b) - f(a) \ . \tag{5.9}$$

Let $A : [a,b] \to \mathcal{B}(\mathcal{H})$ be a $\mathcal{B}(\mathcal{H})$-valued function. One may define an operator $\int_a^b A(s)\, ds$ by

$$\left(\int_a^b A(s)\, ds\right)f = \int_a^b A(s)\, f\, ds \, . \tag{5.10}$$

Its domain is the set of $f \in \mathcal{H}$ for which the latter integral exists. One has similar results to Proposition 5.1 :

<u>PROPOSITION 5.2</u> : Let a,b be finite or infinite.

(a) $\left\| \int_a^b A(s)\, ds \right\| \le \int_a^b \|A(s)\| ds .$ (5.11)

(b) If $B \in \mathcal{B}(\mathcal{H})$, then $B \int_a^b A(s)\, ds = \int_a^b BA(s)\, ds$. (5.12)

(c) If $\{A(s)\}$ is strongly continuous and $(a,b]$ is finite, then $\int_a^b A(s)\, ds$ exists and is in $\mathcal{B}(\mathcal{H})$. If $(a,b]$ is infinite the integral exists and is in $\mathcal{B}(\mathcal{H})$ provided that also $\int_a^b \|A(s)\|\, ds < \infty$. (5.13)

PROPOSITION 5.3 : If $A = A^*$, then

$$(A-z)^{-1} = \begin{cases} i \int_0^\infty e^{izs}\, e^{-iAs}\, ds & \text{if} \quad \operatorname{Im} z > 0 \\ -i \int_{-\infty}^0 e^{izs}\, e^{-iAs}\, ds & \text{if} \quad \operatorname{Im} z < 0 \end{cases} .$$

Proof : Let $\operatorname{Im} z > 0$. Since $e^{izs}\, e^{-iAs}$ is strongly continuous in s and $\|e^{izs}\, e^{-iAs}\, f\| = e^{-(\operatorname{Im} z)s}\, \|f\|$, the function $s \mapsto e^{izs}\, e^{-iAs}\, f$ is strongly Riemann integrable on $[0,\infty)$ for each $f \in \mathcal{H}$, i.e. $i \int_0^\infty ds\, e^{izs}\, e^{-iAs}\, f$ defines a vector in $\mathcal{H}$.

From the functional calculus and Fubini's Theorem one has for all $f,g \in \mathcal{H}$:

$$(g,(A-z)^{-1}f) = \int_{\mathbb{R}} \frac{1}{\lambda-z}\, d(g,E_\lambda f) = \int_{\mathbb{R}} [i \int_0^\infty e^{i(z-\lambda)s}\, ds]\, d(g,E_\lambda f)$$
(5.14)
$$= i \int_0^\infty ds\, e^{izs} \int_{\mathbb{R}} e^{-i\lambda s}\, d(g,E_\lambda f) = i \int_0^\infty ds\, e^{izs}(g,e^{-iAs}f).$$

Since the integral below defines a vector in $\mathcal{H}$, it follows from the Riesz representation theorem that

$$(g,(A-z)^{-1}f) = (g,i\int_0^{\infty} ds\, e^{izs}\, e^{-iAs}\, f)\ . \blacksquare$$

Now let $\{E_\lambda\}$ be a spectral family, $[a,b]$ a finite interval and $\Pi = \{\lambda_0,\ldots,\lambda_n\ ;\ \mu_1,\ldots,\mu_n\}$ be a partition of $[a,b]$. Let $A,B : [a,b] \to \mathcal{B}(\mathcal{H})$ and $f \in \mathcal{H}$. Define

$$\int_a^b A_\lambda dE_\lambda B_\lambda f = \underset{|\Pi|\to 0}{\text{s-lim}} \sum_{k=1}^{n} A(\mu_k)\, E_{(\lambda_{k-1},\lambda_k]} B(\mu_k) f \tag{5.15}$$

if the limit exists and is independent of the chosen sequence of partitions. This defines a linear operator called the *spectral integral* $\int_a^b A_\lambda dE_\lambda B_\lambda$, whose domain is the set of all $f \in \mathcal{H}$ for which the above limit exists.

A sufficient condition for the existence e.g. of $\int_a^b A_\lambda dE_\lambda$ is the Hölder continuity of $\{A_\lambda\}$ in norm with index $\alpha > \frac{1}{2}$, i.e.

$$\|A_\lambda - A_\mu\| \le M|\lambda-\mu|^\alpha \quad \text{for all} \quad \lambda,\mu \in [a,b]\ . \tag{5.16}$$

We shall not need such results.

Improper spectral integrals are defined as strong limits, e.g.

$$\int_a^\infty A_\lambda dE_\lambda B_\lambda = \underset{b\to\infty}{\text{s-lim}} \int_a^b A_\lambda dE_\lambda B_\lambda\ , \tag{5.17}$$

if the limit exists.

<u>LEMMA 5.4</u> : Let $A = A^*$, $\{E_\lambda\}$ the spectral family of A, $[a,b]$ finite or infinite and $B : [a,b] \to \mathcal{B}(\mathcal{H})$ such that $\|B(\lambda)\| \le M < \infty$ for all $\lambda \in [a,b]$. Then

$$\int_a^b B(\lambda)(A-\lambda)dE_\lambda = 0\ . \tag{5.18}$$

Proof : It suffices to assume $[a,b]$ to be a finite interval. Let $\Pi = \{\lambda_k \, ; \, \mu_k\}$ be a partition of $[a,b]$ and $\Delta_k = (\lambda_{k-1}, \lambda_k]$, $k = 1,\ldots,n$.

One has

$$\|(A-\mu_k)\, E_{\Delta_k} f\|^2 = \int_{\lambda_{k-1}}^{\lambda_k} (\lambda-\mu_k)^2 \; d\|E_\lambda E_{\Delta_k} f\|^2 \tag{5.19}$$

$$\leq (\lambda_k-\lambda_{k-1})^2 \, \|E_{\Delta_k} f\|^2 \quad .$$

Then, by the triangle inequality, (5.19), and Schwarz inequality, we get

$$\|\sum_{k=1}^{n} B(\mu_k)(A-\mu_k)E_{\Delta_k} f\| \leq M \sum_{k=1}^{n} \|(A-\mu_k)E_{\Delta_k} f\|$$

$$\leq M \sum_{k=1}^{n} |\Pi|^{\frac{1}{2}} (\lambda_k-\lambda_{k-1})^{\frac{1}{2}} \|E_{\Delta_k} f\|$$

$$\leq M \, |\Pi|^{\frac{1}{2}} \{\sum_{k=1}^{n} (\lambda_k-\lambda_{k-1})\}^{\frac{1}{2}} \{\sum_{k=1}^{n} \|E_{\Delta_k} f\|^2\}^{\frac{1}{2}} \quad .$$

Since $(E_{\Delta_k} f, E_{\Delta_j} f) = 0$ if $k \neq j$ one has

$$\sum_{k=1}^{n} \|E_{\Delta_k} f\|^2 = \|\sum_{k=1}^{n} E_{\Delta_k} f\|^2 = \|E_{(a,b]} f\|^2 \; , \tag{5.20}$$

and

$$\|\sum_{k=1}^{n} B(\mu_k)(A-\mu_k)E_{\Delta_k} f\| \leq M \, |\Pi|^{\frac{1}{2}} \, (b-a)^{\frac{1}{2}} \, \|E_{(a,b]} f\| \; ,$$

which converges to zero as $|\Pi| \to 0$. ∎

LEMMA 5.5 : Let A, $\{E_\lambda\}$ be as above. Let $\phi : \mathbb{R} \to \mathbb{C}$ be a continuous and bounded function. Then $\phi(A)$, defined by the functional calculus, is also given as a spectral integral :

$$\phi(A) = \int_{\mathbb{R}} \phi(\lambda)\, dE_\lambda \ . \tag{5.21}$$

Proof : We use the same notation as above. Let $\Delta = (a,b]$ be finite, $f \in \mathcal{H}$ and set $f_\Delta = E_{(a,b]}f$. By the functional calculus

$$\phi(A)f_\Delta = \underset{|\Pi|\to 0}{\text{w-lim}} \sum_{k=1}^{n} \phi(\mu_k)E_{\Delta_k}f_\Delta \equiv \underset{|\Pi|\to 0}{\text{w-lim}} \sum\nolimits_\Pi(f) \ .$$

As in (5.20) above

$$\left\| \sum_{k=1}^{n} \phi(\mu_k)E_{\Delta_k}f_\Delta \right\|^2 = \sum_{k=1}^{n} |\phi(\mu_k)|^2 \, \| E_{\Delta_k}f_\Delta \|^2 \ . \tag{5.22}$$

By the functional calculus and Proposition 4.5, this converges as $|\Pi| \to 0$ to

$$\int_a^b |\phi(\lambda)|^2 \, d\|E_\lambda f_\Delta\|^2 = (f_\Delta, \phi(A)^*\phi(A)f_\Delta) \ .$$

So $\|\sum_\Pi(f)\|$ converges to $\|\phi(A)f_\Delta\|$ as $|\Pi| \to 0$. Hence by Lemma 2.1, $\sum_\Pi(f)$ converges strongly to $\phi(A)f_\Delta$.

Now let $a \to -\infty$, $b \to +\infty$. Then $\phi(A)f - \phi(A)f_\Delta = (I-E_{(a,b]})\phi(A)f$ converges strongly to zero, i.e.

$$\phi(A)f = \underset{\substack{a\to\infty \\ b\to-\infty}}{\text{s-lim}} \int_a^b \phi(\lambda)\, dE_\lambda f \ . \ \blacksquare$$

PROPOSITION 5.6 : Let $A = A^*$, $B = B^*$ and let $\{E_\lambda\}$ be the spectral family of A. Let $T \in \mathcal{B}(\mathcal{H})$ and $\varepsilon > 0$. Then the integrals below exist, and one has the identities

$$\int_{-\infty}^{0} dt\, e^{\varepsilon t}\, e^{iBt}\, T\, e^{-iAt} = -i \int_{\mathbb{R}} (B-\lambda-i\varepsilon)^{-1}\, T\, dE_\lambda \tag{5.23}$$

$$\int_{0}^{\infty} dt\, e^{-\varepsilon t}\, e^{iBt}\, T\, e^{-iAt} = i \int_{\mathbb{R}} (B-\lambda+i\varepsilon)^{-1}\, T\, dE_\lambda . \tag{5.24}$$

Proof :

(i) Since $\|e^{-\varepsilon t}\, e^{iBt}\, T\, e^{-iAt}\| \leq e^{-\varepsilon t}\, \|T\|$ is integrable over $[0,\infty)$ and $t \mapsto e^{-\varepsilon t}\, e^{iBt}\, T\, e^{-iAt}$ is strongly continuous, $J_+ \equiv \int_0^\infty dt\, e^{-\varepsilon t}\, e^{iBt}\, T\, e^{-iAt}$ exists by Proposition 5.2 (c).

(ii) Let $a > 0$, $\Delta = (-a,a]$ and $\Pi = \{\lambda_k\, ;\, \mu_k\}$ a partition of Δ. Let $\Delta_k = (\lambda_{k-1},\lambda_k]$. For $f \in \mathcal{H}$, define

$$J_\Pi(f) = \int_0^\infty dt\, e^{-\varepsilon t}\, e^{iBt}\, T \sum_{k=1}^{n} e^{-i\mu_k t}\, E_{\Delta_k}\, f \quad .$$

Then

$$\|J_+ E_\Delta f - J_\Pi(f)\| \leq \|T\| \int_0^\infty dt\, e^{-\varepsilon t}\, \|e^{-iAt}\, E_\Delta f - \sum_{k=1}^{n} e^{-i\mu_k t}\, E_{\Delta_k}\, E_\Delta f\| \; .$$

The integrand is majorized uniformly in Π by the integrable function $2e^{-\varepsilon t}\, \|f\|$ and converges pointwise to zero as $|\Pi| \to 0$ by Lemma 5.5. Thus by the Lebesgue dominated convergence Theorem :

$$\underset{|\Pi| \to 0}{\text{s-lim}}\; J_\Pi(f) = J_+\, E_\Delta\, f \quad . \tag{5.25}$$

(iii) Now, setting $t = -\tau$ and using Proposition 5.3, one has

$$J_\Pi(f) = \sum_{k=1}^{n} \int_{-\infty}^{0} d\tau \, e^{i(\mu_k - i\varepsilon)\tau} \, e^{-iB\tau} \, T \, E_{\Delta_k} f$$

$$= i \sum_{k=1}^{n} (B - \mu_k + i\varepsilon)^{-1} \, T \, E_{\Delta_k} f \, .$$

By (5.25) : $J_+ E_\Delta f = i \int_{-a}^{a} (B - \lambda + i\varepsilon)^{-1} \, T \, dE_\lambda f$ and the spectral integral exists. As $a \to \infty$, E_Δ converges strongly to I. Hence

$$J_+ f = \text{s-}\lim_{a \to \infty} J_+ E_{(-a,a]} f = i \int_{\mathbb{R}} (B - \lambda + i\varepsilon)^{-1} \, T \, dE_\lambda f \quad ,$$

showing also the existence of the spectral integral.

The other formula is proved similarly. ∎

<u>PROPOSITION 5.7</u> : Let $A = A^*$ and set $R_z = (A-z)^{-1}$. Then

$$E_{(a,b]} = \text{s-}\lim_{\delta \to +0} \text{s-}\lim_{\varepsilon \to +0} \frac{1}{2\pi i} \int_{a+\delta}^{b+\delta} d\lambda [R_{\lambda+i\varepsilon} - R_{\lambda-i\varepsilon}] \, .$$

If $a,b \notin \sigma_p(A)$, then

$$E_{(a,b)} = E_{[a,b]} = E_{(a,b]} = \text{s-}\lim_{\varepsilon \to +0} \frac{1}{2\pi i} \int_{a}^{b} d\lambda [R_{\lambda+i\varepsilon} - R_{\lambda-i\varepsilon}]. \quad (5.26)$$

<u>Proof</u> : Let $\delta, \varepsilon > 0$.

Set

$$\psi_{\delta,\varepsilon}(\lambda) = \frac{1}{2\pi i} \int_{a+\delta}^{b+\delta} \left[\frac{1}{\mu-\lambda-i\varepsilon} - \frac{1}{\mu-\lambda+i\varepsilon}\right] d\mu$$

$$= \frac{1}{2\pi i} \int_{a+\delta}^{b+\delta} \frac{2i\varepsilon}{(\mu-\lambda)^2+\varepsilon^2} \, d\mu = \frac{1}{\pi} \left\{\arctan \frac{b+\delta-\lambda}{\varepsilon} - \arctan \frac{a+\delta-\lambda}{\varepsilon}\right\}.$$

$\psi_{\delta,\varepsilon}$ is continuous, $|\psi_{\delta,\varepsilon}(\lambda)| \le 1$ and

$$\lim_{\delta\to+0} \lim_{\varepsilon\to+0} \psi_{\delta,\varepsilon}(\lambda) = \begin{cases} 1 & \text{if } \lambda \in (a,b] \\ 0 & \text{if } \lambda \notin (a,b] . \end{cases}$$

As in the proof of Lemma 5.5 [relation (5.22)] :

$$\left\| \int_{(a,b]} [1 - \psi_{\delta,\varepsilon}(\lambda)]\, dE_\lambda f \right\|^2 = \int_{(a,b]} |1 - \psi_{\delta,\varepsilon}(\lambda)|^2\, d(f,E_\lambda f). \quad (5.27)$$

The last integrand is bounded by 4 uniformly in δ,ε and converges pointwise to zero on $(a,b]$ as $\delta,\varepsilon \to +0$. By the Lebesgue dominated convergence theorem :

$$\underset{\delta\to+0}{\text{s-lim}}\ \underset{\varepsilon\to+0}{\text{s-lim}} \int_{(a,b]} [1 - \psi_{\delta,\varepsilon}(\lambda)]\, dE_\lambda f = 0 . \quad (5.28)$$

Similarly

$$\underset{\delta\to+0}{\text{s-lim}}\ \underset{\varepsilon\to+0}{\text{s-lim}} \int_{\mathbb{R}\setminus(a,b]} \psi_{\delta,\varepsilon}(\lambda)\, dE_\lambda f = 0 . \quad (5.29)$$

If one adds (5.28) and (5.29) one obtains

$$E_{(a,b]} = \int_{a+0}^{b+0} dE_\lambda = \underset{\delta\to+0}{\text{s-lim}}\ \underset{\varepsilon\to+0}{\text{s-lim}} \int_{\mathbb{R}} \psi_{\delta,\varepsilon}(\lambda)\, dE_\lambda . \quad (5.30)$$

Hence, by Fubini's theorem, one gets

$$\int_{\mathbb{R}} \psi_{\delta,\varepsilon}(\lambda)\, dE_\lambda = \frac{1}{2\pi i} \int_{\mathbb{R}} \{\int_{a+\delta}^{b+\delta} d\mu\, [\frac{1}{\mu-\lambda-i\varepsilon} - \frac{1}{\mu-\lambda+i\varepsilon}]\}\, dE_\lambda$$

$$= \frac{1}{2\pi i} \int_{a+\delta}^{b+\delta} d\mu \int_{\mathbb{R}} [\frac{1}{\mu-\lambda-i\varepsilon} - \frac{1}{\mu-\lambda+i\varepsilon}]\, dE_\lambda$$

$$= \frac{1}{2\pi i} \int_{a+\delta}^{b+\delta} d\mu\, [-R_{\mu-i\varepsilon} + R_{\mu+i\varepsilon}]\ ,$$

which together with (5.30) proves the first part. The proof of the second part is similar and left as an exercise. ∎

6 Wave operators

Let H_o and H be two self-adjoint operators. The *wave operators* $\Omega_\pm$, also denoted $\Omega_\pm(H,H_o)$, for the pair $\{H,H_o\}$ are defined as

$$\Omega_\pm \equiv \Omega_\pm(H,H_o) = \underset{t\to\pm\infty}{\text{s-lim}}\; e^{iHt} e^{-iH_o t} E_{ac}(H_o) , \tag{6.1}$$

if these limits exist. We first assume that $\Omega_\pm$ exist and derive some of their properties, and later we shall prove their existence for certain Schrödinger operators.

PROPOSITION 6.1 :

(a) $\Omega_\pm$ are partial isometries with initial set $\mathcal{H}_{ac}(H_o)$. In particular

$$\Omega_\pm^*\Omega_\pm = E_{ac}(H_o),\ \Omega_\pm E_{ac}(H_o) = \Omega_\pm,\ E_{ac}(H_o)\Omega_\pm^* = \Omega_\pm^*. \tag{6.2}$$

(b) Define $F_\pm = \Omega_\pm\Omega_\pm^*$. $F_\pm$ is the projection onto the range of $\Omega_\pm$ respectively. In particular

$$F_\pm\Omega_\pm = \Omega_\pm,\ \Omega_\pm^* F_\pm = \Omega_\pm. \tag{6.3}$$

(c)
$$\Omega_\pm^* = \underset{t\to\pm\infty}{\text{s-lim}}\; e^{iH_o t} e^{-iHt} F_\pm . \tag{6.4}$$

Remark : A *partial isometry* Ω is an operator that maps a subspace M isometrically onto another subspace M', and which is zero on $M^\perp$. M is called the *initial set* of Ω, M' its *range*.

Proof :

(a) Clearly $\Omega f = 0$ if $f \perp \mathcal{H}_{ac}(H_o)$. If $h \in \mathcal{H}_{ac}(H_o)$, then by Lemma 2.1 and the unitarity of e^{iHt}, $e^{-iH_o t}$:

$$\|\Omega_{\pm} h\| = \lim_{t \to \pm\infty} \|e^{iHt} e^{-iH_o t} h\| = \|h\|.$$

If $f,g \in \mathcal{H}_{ac}(H_o)$, then $f+\alpha g \in \mathcal{H}_{ac}(H_o)$ for any $\alpha \in \mathbb{C}$, so that by polarization identity

$$(\Omega_{\pm} f,\Omega_{\pm} g) = \frac{1}{4}\{\|\Omega_{\pm}(f+g)\|^2 - \|\Omega_{\pm}(f-g)\|^2 - i\|\Omega_{\pm}(f+ig)\|^2 + i\|\Omega_{\pm}(f-ig)\|^2\}$$

$$= \frac{1}{4}\{\|f+g\|^2 - \|f-g\|^2 - i\|f+ig\|^2 + i\|f-ig\|^2\} = (f,g) .$$

Thus $\Omega_{\pm}$ are isometric on $\mathcal{H}_{ac}(H_o)$.

If $f,g \in \mathcal{H}$, then as above

$$\begin{aligned}(\Omega_{\pm}^{*}\Omega_{\pm} f,g) = (\Omega_{\pm} f,\Omega_{\pm} g) &= (\Omega_{\pm} E_{ac}(H_o)f,\Omega_{\pm} E_{ac}(H_o)g)\\ &= (E_{ac}(H_o)f,E_{ac}(H_o)g)\\ &= (E_{ac}(H_o)f,g) .\end{aligned}$$

Hence $\Omega_{\pm}^{*}\Omega_{\pm} = E_{ac}(H_o)$. Clearly $\Omega_{\pm} E_{ac}(H_o) = \Omega_{\pm}$, and the last identity follows from this by taking the adjoint.

(b) $$F_{\pm}^{*} = (\Omega_{\pm}\Omega_{\pm}^{*})^{*} = \Omega_{\pm}^{**}\Omega_{\pm}^{*} = \Omega_{\pm}\Omega_{\pm}^{*} = F_{\pm} .$$

$$F_{\pm}^{2} = \Omega_{\pm}\Omega_{\pm}^{*}\Omega_{\pm}\Omega_{\pm}^{*} = \Omega_{\pm} E_{ac}(H_o)\Omega_{\pm}^{*} = \Omega_{\pm}\Omega_{\pm}^{*} = F_{\pm} .$$

Hence $F_\pm$ are projections. Clearly (range $F_\pm$) $\subseteq$ (range $\Omega_\pm$). If $g = \Omega_\pm f \in$ (range $\Omega_\pm$), then

$$F_\pm g = \Omega_\pm \Omega_\pm^* \Omega_\pm f = \Omega_\pm E_{ac}(H_0)f = \Omega_\pm f = g \ .$$

Hence $g \in$ (range $F_\pm$). Therefore (range $F_\pm$) = (range $\Omega_\pm$).

(c) Let $f \in \mathcal{H}$. Then

$$\|e^{iH_0 t} e^{-iHt} \Omega_\pm f - E_{ac}(H_0)f\| = \|\Omega_\pm f - e^{iHt} e^{-iH_0 t} E_{ac}(H_0)f\| ,$$

which converges to zero as $t \to \pm\infty$, i.e. $\operatorname*{s-lim}_{t\to\pm\infty} e^{iH_0 t} e^{-iHt}\Omega_\pm = E_{ac}(H_0)$.

Multiplying by $\Omega_\pm^*$ on the right and using (a), one obtains

$$\operatorname*{s-lim}_{t\to\pm\infty} e^{iH_0 t} e^{-iHt} F_\pm = E_{ac}(H_0)\Omega_\pm^* = \Omega_\pm^* . \ \blacksquare$$

<u>PROPOSITION 6.2</u> :

(a) $F_\pm \mathcal{H} \subseteq \mathcal{H}_{ac}(H)$. (6.5)

(b) $\Omega_\pm$ intertwine H and H_0, i.e.

$$\phi(H)\Omega_\pm = \Omega_\pm \phi(H_0) \qquad (6.6)$$

for each bounded continuous function $\phi : \mathbb{R} \to \mathbb{C}$. Also

$$(H-z)^{-1}\Omega_\pm = \Omega_\pm (H_0 - z)^{-1} \qquad \forall z \in \rho(H) \cap \rho(H_0) \qquad (6.7)$$

$$e^{iHt}\Omega_\pm = \Omega_\pm e^{iH_o t} \qquad \forall t \in \mathbb{R} \tag{6.8}$$

$$E_\lambda \Omega_\pm = \Omega_\pm E^0_\lambda \qquad \forall \lambda \in \mathbb{R} \ , \tag{6.9}$$

where $\{E_\lambda\}$ and $\{E^0_\lambda\}$ are the spectral families of H and H_o respectively.

Proof :

(i) We first prove (6.8) :

$$\begin{aligned} e^{iHt}\Omega_\pm &= \underset{\tau \to \pm\infty}{\text{s-lim}}\, e^{iHt}\, e^{iH\tau}\, e^{-iH_o\tau}\, E_{ac}(H_o) \\ &= \underset{\sigma \to \pm\infty}{\text{s-lim}}\, e^{iH\sigma}\, e^{-iH_o\sigma}\, e^{iH_o t}\, E_{ac}(H_o) \ . \end{aligned} \tag{6.10}$$

Since $e^{iH_o t} E_{ac}(H_o) = E_{ac}(H_o)\, e^{iH_o t}$, this gives (6.8).

(ii) Now for Im z > 0 [see Proposition 5.3] :

$$\begin{aligned} (H-z)^{-1}\Omega_\pm &= i\int_0^\infty e^{izs}\, e^{-iHs}\, \Omega_\pm\, ds = i\int_0^\infty e^{izs}\, \Omega_\pm\, e^{-iH_o s}\, ds \\ &= \Omega_\pm (H_o - z)^{-1} \ . \end{aligned}$$

Similarly one gets (6.7) for Im z < 0. If z is real, (6.7) holds by analytic continuation.

(iii) (6.9) follows from (6.7), Proposition 5.7 and (6.2) :

$$2\pi i \; E_{(a,b]}\Omega_\pm = \underset{\delta\to+0}{\text{s-lim}}\;\underset{\varepsilon\to+0}{\text{s-lim}}\int_{a+\delta}^{b+\delta} d\lambda[(H-\lambda-i\varepsilon)^{-1} - (H-\lambda+i\varepsilon)^{-1}]\;\Omega_\pm$$

$$= \underset{\delta\to+0}{\text{s-lim}}\;\underset{\varepsilon\to+0}{\text{s-lim}}\int_{a+\delta}^{b+\delta} \Omega_\pm[(H_0-\lambda-i\varepsilon)^{-1} - (H_0-\lambda+i\varepsilon)^{-1}]\;d\lambda$$

$$= 2\pi i\;\Omega_\pm\;E^o_{(a,b]}\;.$$

Hence

$$E_b\Omega_\pm = \underset{a\to-\infty}{\text{s-lim}}\;E_{(a,b]}\Omega_\pm = \underset{a\to-\infty}{\text{s-lim}}\;\Omega_\pm\;E^o_{(a,b]} = \Omega_\pm\Omega^0_b\;.$$

(iv) (6.6) follows from (6.9) and the functional calculus.

(v) Let $f \in F_+\mathcal{H}$, i.e. $f = \Omega_+ g$ with $g = \Omega_+^* f \in \mathcal{H}_{ac}(H_0)$.

Then by (6.9) one has :

$$m_f(\Delta) \equiv (f,E_\Delta f) = (g,\Omega_+^* E_\Delta \Omega_+ g) = (g,E^0_\Delta \Omega_+^* \Omega_+ g) = (g,E^0_\Delta g) = m^o_g(\Delta)\;.$$

Since $g \in \mathcal{H}_{ac}(H_0)$, $m^o_g(.)$ is an absolutely continuous measure. Hence $m_f(.)$ is absolutely continuous, in other words $f \in \mathcal{H}_{ac}(H)$. This proves (a). ∎

Proposition 6.2 (a) implies that, if $\sigma_{ac}(H_0) \neq \phi$ and Ω_+ or Ω_- exists, then $\sigma_{ac}(H) \neq \phi$.

<u>COROLLARY 6.3</u> : One has in the notation of Proposition 6.2 :

$$\Omega_\pm^*\phi(H) = \phi(H_0)\;\Omega_\pm^* \qquad (6.11)$$

$$\Omega_\pm^*\;(H-z)^{-1} = (H_0-z)^{-1}\;\Omega_\pm^* \qquad (6.12)$$

$$\Omega_\pm^* e^{iHt} = e^{iH_0 t} \Omega_\pm^* \tag{6.13}$$

$$\Omega_\pm^* E_\lambda = E_\lambda^0 \Omega_\pm^* \tag{6.14}$$

$$F_\pm e^{iHt} = e^{iHt} F_\pm . \tag{6.15}$$

Proof : [(6.11) - (6.14)] follow from the corresponding statements in Proposition 6.2 by taking the adjoint, e.g.

$$\Omega_\pm^*(H-z)^{-1} = [(H-\bar{z})^{-1}\Omega_\pm]^* = [\Omega_\pm(H_0-\bar{z})^{-1}]^* = (H_0-z)^{-1} \Omega_\pm^*.$$

(6.15) is obtained as follows by using (6.13) and (6.8) :

$$F_+ e^{iHt} = \Omega_+ \Omega_+^* e^{iHt} = \Omega_+ e^{iH_0 t} \Omega_+^* = e^{iHt} \Omega_+ \Omega_+^* = e^{iHt} F_+ . \blacksquare$$

PROPOSITION 6.4 : One has

$$\Omega_+ = \underset{\varepsilon\to+0}{\text{s-lim}}\ \varepsilon \int_0^\infty dt\ e^{-\varepsilon t} e^{iHt} e^{-iH_0 t} E_{ac}(H_o)$$

$$= \underset{\varepsilon\to+0}{\text{s-lim}}(i\varepsilon) \int_{\mathbb{R}} (H-\lambda+i\varepsilon)^{-1} dE_\lambda^0 E_{ac}(H_o) .$$

$$\Omega_- = \underset{\varepsilon\to+0}{\text{s-lim}}\ \varepsilon \int_{-\infty}^0 dt\ e^{\varepsilon t} e^{iHt} e^{-iH_0 t} E_{ac}(H_o)$$

$$= \underset{\varepsilon\to+0}{\text{s-lim}}\ (-i\varepsilon) \int_{\mathbb{R}} (H-\lambda-i\varepsilon)^{-1} dE_\lambda^0 E_{ac}(H_o) .$$

$$\Omega_+^* = \underset{\varepsilon\to+0}{\text{s-lim}}\ \varepsilon \int_0^\infty dt\ e^{-\varepsilon t} e^{iH_0 t} e^{-iHt} F_+$$

$$= \underset{\varepsilon\to+0}{\text{s-lim}}\ (i\varepsilon) \int_{\mathbb{R}} (H_0-\lambda+i\varepsilon)^{-1} dE_\lambda F_+ .$$

$$\Omega_-^* = \underset{\varepsilon\to+0}{\text{s-lim}}\ \varepsilon \int_{-\infty}^{0} dt\ e^{\varepsilon t}\ e^{iH_0 t}\ e^{-iHt}\ F_-$$

$$= \underset{\varepsilon\to+0}{\text{s-lim}}\ (-i\varepsilon) \int_{\mathbb{R}} (H_0-\lambda-i\varepsilon)^{-1}\ dE_\lambda\ F_- \ .$$

Proof : The first identity for Ω_+ is true because the existence of the limit $\underset{t\to+\infty}{\text{s-lim}}\ e^{iHt}\ e^{-iH_0 t}\ E_{ac}(H_0)$ implies the existence of the Abel limit (the proof is standard). The second identity for Ω_+ follows immediately from Proposition 5.6.

Similarly, since $\Omega_+^* = \underset{t\to+\infty}{\text{s-lim}}\ e^{iH_0 t}\ e^{-iHt}\ F_+$, one obtains the identities for Ω_+^*. Same argument for Ω_- and Ω_-^*. ■

Remark : The study of the limits $\underset{t\to\pm\infty}{\text{s-lim}}\ e^{iHt}\ e^{-iH_0 t}\ E_{ac}(H_0)$ is called *time-dependent scattering theory*, since physically the parameter t corresponds to the time. The study of the operators in the last column in Proposition 6.4 is called *stationary state scattering theory*. It is possible that the stationary wave operators (i.e. the limits in Proposition 6.4) exist without the usual time-dependent ones existing.

PROPOSITION 6.5 : (*Lippmann-Schwinger equations*) : [28]

Suppose $H = H_0 + V$, i.e. $D(H) = D(H_0) \subseteq D(V)$. Then

$$\Omega_\pm = E_{ac}(H_o) - \underset{\varepsilon\to+0}{\text{s-lim}} \int_{\mathbb{R}} (H-\lambda\pm i\varepsilon)^{-1}\ V\ dE_\lambda^0\ E_{ac}(H_0) \qquad (6.16)$$

$$= E_{ac}(H_o) - \underset{\varepsilon\to+0}{\text{s-lim}} \int_{\mathbb{R}} (H_0-\lambda\pm i\varepsilon)^{-1}\ V\ \Omega_\pm\ dE_\lambda^0 \ . \qquad (6.17)$$

Proof : Since $\|(H-\lambda\pm i\varepsilon)^{-1}\| \le \frac{1}{\varepsilon}$, we have by Lemma 5.4

$$\int (H-\lambda\pm i\varepsilon)^{-1}\ (H_0-\lambda)\ dE_\lambda^0 = 0 \ . \qquad (6.18)$$

Let $0 < a < \infty$ and $\Delta = (-a,a]$. Then $E_{\Delta}^{o}\mathcal{H} \subseteq D(H_o)$, hence

$$HE_{\Delta}^{o} = H_oE_{\Delta}^{o} + VE_{\Delta}^{o} \quad .$$

By using (6.18) one obtains

$$(i\varepsilon) \int_{\mathbb{R}} (H-\lambda+i\varepsilon)^{-1} E_{\Delta}^{o}\, dE_{\lambda}^{o}\, E_{ac}(H_o)$$

$$= \int_{\mathbb{R}} (H-\lambda+i\varepsilon)^{-1} (H_o-\lambda+i\varepsilon)\, E_{\Delta}^{o}\, dE_{\lambda}^{o}\, E_{ac}(H_o)$$

$$= \int_{\mathbb{R}} (H-\lambda+i\varepsilon)^{-1} [(H-\lambda+i\varepsilon)-V]\, E_{\Delta}^{o}\, dE_{\lambda}^{o}\, E_{ac}(H_o)$$

$$= E_{\Delta}^{o}\, E_{ac}(H_o) - \int_{-a}^{a} (H-\lambda+i\varepsilon)^{-1}\, V\, dE_{\lambda}^{o}\, E_{ac}(H_o) \ .$$

As $a \to \infty$, E_{Δ}^{o} converges strongly to I and the first integral converges by Proposition 5.6. Hence the last integral converges and

$$(i\varepsilon) \int_{\mathbb{R}} (H-\lambda+i\varepsilon)^{-1}\, dE_{\lambda}^{o}\, E_{ac}(H_o) = E_{ac}(H_o) - \int_{\mathbb{R}} (H-\lambda+i\varepsilon)^{-1}\, V\, dE_{\lambda}^{o}\, E_{ac}(H_o).$$

As $\varepsilon \to +0$, the left-hand side converges to Ω_+ by Proposition 6.4, proving (6.16). (6.17) is obtained similarly. ∎

(6.17) gives an integral equation for $\Omega_{\pm}$ involving operators depending on H_o and the perturbation V.

<u>PROPOSITION 6.6</u> : If $f \in D(H_o)$, then $\Omega_{\pm} f \in D(H)$ and $H\, \Omega_{\pm}\, f = \Omega_{\pm}\, H_o\, f$. If $g \in D(H)$, then $\Omega_{\pm}^{*}\, g \in D(H_o)$ and $H_o\, \Omega_{\pm}^{*}\, g = \Omega_{\pm}^{*}\, Hg$.

<u>Proof</u> : By Proposition 6.2 and since $\|\Omega_{\pm}\| = 1$, one has

(i) $$\|H\Omega_\pm f\|^2 = \int_{\mathbb{R}} \lambda^2 \, d\|E_\lambda \Omega_\pm f\|^2 = \int_{\mathbb{R}} \lambda^2 \, d\|\Omega_\pm E^0_\lambda f\|^2$$

$$\leq \int \lambda^2 \, d\|E^0_\lambda f\|^2 = \|H_0 f\|^2 < \infty.$$

Hence $\Omega_\pm f \in D(H)$.

(ii) One has

$$\|[\frac{i}{t}(e^{-iH_0 t} - I) - H_0]f\|^2 = \int_{\mathbb{R}} |\frac{i}{t}(e^{-i\lambda t} - 1) - \lambda|^2 d\|E^0_\lambda f\|^2 . \quad (6.19)$$

Now

$$|\frac{i}{t}(e^{-i\lambda t} - 1) - \lambda| = |\int_0^\lambda (e^{-i\mu t} - 1) d\mu| \leq 2|\lambda|.$$

Thus the integrand in (6.19) is majorized uniformly in t by an integrable function and it converges pointwise to zero as $t \to 0$. By the Lebesgue dominated convergence theorem :

$$\operatorname*{s-lim}_{t \to 0} [\frac{i}{t}(e^{-iH_0 t} - I)]f = H_0 f .$$

Similarly $\text{s-lim}\, \frac{i}{t}(e^{-iHt} - I)\Omega_\pm f = H\Omega_\pm f$. Since $\frac{i}{t}(e^{-iHt} - I)\Omega_\pm f =$ $\Omega_\pm [\frac{i}{t}(e^{-iH_0 t} - I)]f$ by Proposition 6.2, we have $H\Omega_\pm f = \Omega_\pm H_0 f$.

The proof of the second part is similar. ∎

PROPOSITION 6.7 : If Ω_+ or Ω_- exists, then $\sigma_{ac}(H) \supseteq \sigma_{ac}(H_0)$. If $F_+ = E_{ac}(H)$ or $F_- = E_{ac}(H)$, then $\sigma_{ac}(H) = \sigma_{ac}(H_0)$.

Proof : *(for the + sign)* :

(i) By (6.15) F_+ commutes with e^{-iHt}, hence with E_λ, hence with H. Since $F_+\mathcal{H} \subseteq \mathcal{H}_{ac}(H)$ [Proposition 6.2(a)], HF_+ is absolutely continuous, and $\sigma(HF_+) = \sigma_{ac}(HF_+) \subseteq \sigma_{ac}(H)$. Thus it suffices to show that

$$\sigma(HF_+) = \sigma_{ac}(H_0) = \sigma(H_0\ E_{ac}(H_0)) \ , \tag{6.20}$$

or equivalently that $\rho(HF_+) = \rho(H_0\ E_{ac}(H_0))$, i.e.

$$(HF_+ - z)^{-1} \in \mathcal{B}(F_+\mathcal{H}) \Leftrightarrow (H_0\ E_{ac}(H_0) - z)^{-1} \in \mathcal{B}(\mathcal{H}_{ac}(H_0)) \ ,$$

where HF_+ is viewed as an operator in $F_+\mathcal{H}$ and $H_0\ E_{ac}(H_0)$ as an operator in $\mathcal{H}_{ac}(H_0)$.

(ii) Suppose $(HF_+ - z)^{-1} \in \mathcal{B}(F_+\mathcal{H})$. Then $(HF_+ - z)^{-1}\ \Omega_+ = (H - z)^{-1}\ \Omega_+$. Let $f \in D(H_0) \cap \mathcal{H}_{ac}(H_0) = D(H_0\ E_{ac}(H_0))$. Then, by using Proposition 6.6, one has

$$\Omega_+^*(HF_+ - z)^{-1}\ \Omega_+\ (H_0\ E_{ac}(H_0) - z)f$$

$$= \Omega_+^*(HF_+ - z)^{-1}(H - z)F_+\ \Omega_+\ f = \Omega_+^*\ \Omega_+\ f = E_{ac}(H_0)f = f \ .$$

Hence $(H_0\ E_{ac}(H_0) - z)$ is invertible on $\mathcal{H}_{ac}(H_0)$, and its inverse is $\Omega_+^*(HF_+ - z)^{-1}\ \Omega_+$ which is in $\mathcal{B}(\mathcal{H}_{ac}(H_0))$.

(iii) The converse implication is obtained similarly by interchanging the roles of Ω_+ and Ω_+^*, of H_0 and H and of $E_{ac}(H_0)$ and F_+. ∎

We now turn to the existence of the wave operators.

LEMMA 6.8 : Let $\mathcal{D}_0$ be a subset of $\mathcal{H}_{ac}(H_0)$ such that the set $\mathcal{D}$ of all finite linear combinations of vectors in $\mathcal{D}_0$ is dense in $\mathcal{H}_{ac}(H_0)$. If $\text{s-lim}_{t\to\pm 0} e^{iHt} e^{-iH_0 t} f$ exists for each $f \in \mathcal{D}_0$, then $\Omega_\pm$ exist.

Proof : The hypothesis implies that $\Omega_\pm f$ exist for $f \in \mathcal{D}$. If $g \in \mathcal{H}_{ac}(H_0)$ and $\varepsilon > 0$, choose $f \in \mathcal{D}$ such that $\|g-f\| < \varepsilon/3$. Then

$$\|e^{iHt} e^{-iH_0 t} g - e^{iHs} e^{-iH_0 s} g\| \le \|e^{iHt} e^{-iH_0 t} f - e^{iHs} e^{-iH_0 s} f\| + 2\|g-f\| ,$$

since $\|e^{iHt}\| = \|e^{iH_0 t}\| = 1$.

Since $\Omega_\pm f$ exists, there exists T such that e.g.

$$\|e^{iHt} e^{-iH_0 t} f - e^{iHs} e^{-iH_0 s} f\| < \frac{\varepsilon}{3} \qquad \text{whenever} \quad t,s > T .$$

Therefore $\{e^{iHt} e^{-iH_0 t} g\}$ is strongly Cauchy, i.e. $\Omega_+ g$ exists. ∎

LEMMA 6.9 : Let $\{W_t\}_{t\ge a}$ be a family of operators and $f \in D(W_t)$ for each t. Suppose that $\{W_t f\}$ is strongly continuously differentiable and that

$$\int_a^\infty d\tau \, \|\frac{d}{d\tau} W_\tau f\| < \infty.$$

Then $\{W_t f\}$ converges strongly as $t \to +\infty$.

Proof : By Proposition 5.2(a), one has

$$\|W_t f - W_s f\| = \|\int_s^t \frac{d}{d\tau} W_\tau f \, d\tau\| \le \int_s^t \|\frac{d}{d\tau} W_\tau f\| \, d\tau$$

which converges to zero as $s,t \to \infty$. Thus $\{W_t f\}$ is strongly Cauchy. ∎

Up to this point we have considered an abstract pair of self-adjoint operators H and H_o. For the remainder of this chapter we consider the case where H_o is equal to the operator K_o introduced in Example 3 of chapter 3. In the same spirit, we shall in later chapters write H_o when dealing with an abstract self-adjoint operator and K_o when treating Schrödinger operators.

<u>LEMMA 6.10</u> : Let $\mathcal{H} = L^2(\mathbb{R}^n)$, $H_o = K_o$ and $f \in L^1(\mathbb{R}^n) \cap L^2(\mathbb{R}^n)$. Then for $t \neq 0$

$$(e^{-iK_o t} f)(x) = (4\pi i t)^{-\frac{n}{2}} \int e^{i\frac{|x-y|^2}{4t}} f(y)\, dy , \qquad (6.21)$$

the branch of the square root being such that

$$(4\pi i t)^{-\frac{n}{2}} = |4\pi t|^{-\frac{n}{2}} \cdot \begin{cases} \exp(-\frac{in\pi}{4}) & \text{if } t > 0 \\ \exp(i\frac{n\pi}{4}) & \text{if } t < 0 . \end{cases}$$

<u>Proof</u> :

(i) We use the following result : if $\alpha \in \mathbb{C}$, Re $\alpha > 0$, the Fourier transform of $\exp(-\alpha \frac{|x|^2}{2})$ is $\alpha^{-\frac{n}{2}} \exp(-\frac{|k|^2}{2\alpha})$ with the above choice of the branch of $\sqrt{z}$.

(ii) Let $a \in \mathbb{R}^n$, $f_a(x) = \exp(-|x-a|^2)$. Then

$$\begin{aligned}(e^{-iK_o t} f)(x) &= (2\pi)^{-\frac{n}{2}} \int dk\, e^{ik.x} e^{-i|k|^2 t} \tilde{f}_a(k) \\ &= (4\pi)^{-\frac{n}{2}} \int dk\, e^{ik.x} e^{-i|k|^2 t} e^{-\frac{|k|^2}{4}} e^{-ik.a} \\ &= (1+4it)^{-\frac{n}{2}} \exp[-\frac{(x-a)^2}{1+4it}] . \end{aligned} \qquad (6.22)$$

An explicit calculation, using (i), shows that

$$(4\pi it)^{-\frac{n}{2}} \int dy\, e^{i\frac{|x-y|^2}{4t}} f_a(y)$$

$$= (1+4it)^{-\frac{n}{2}} \exp[-\frac{|x-a|^2}{1+4it}] .$$

Hence the result of the Lemma holds for $f = f_a$ and for the set $\mathcal{D}$ of all finite linear combinations of the functions f_a. $\mathcal{D}$ is dense in $L^2(\mathbb{R}^n)$ and in $L^1(\mathbb{R}^n)$, so that one can extend the result to $L^1(\mathbb{R}^n) \cap L^2(\mathbb{R}^n)$. ∎

<u>COROLLARY 6.11</u> : Let $f \in L^1(\mathbb{R}^n) \cap L^2(\mathbb{R}^n)$. Let $V(.) \in L^2(\mathbb{R}^n)$ and let V be the maximal multiplication operator by V(x). Then

(a) $$\|e^{-iK_0t} f\|_\infty \equiv \sup_{x\in\mathbb{R}^n} |(e^{-iK_0t} f)(x)| \leq |4\pi t|^{-\frac{n}{2}} \|f\|_1 \tag{6.23}$$

(b) $$\lim_{t\to\pm\infty} \|V e^{-iK_0t} f\| = 0 \tag{6.24}$$

(c) $$\int_1^\infty \|V e^{-iK_0t} f\| \, dt < \infty \quad \text{if} \quad n \geq 3 . \tag{6.25}$$

<u>Proof</u> :

(a) is evident.

(b) $$\|V e^{-iK_0t} f\|^2 = \int dx\, |V(x)|^2 \, |(e^{-iK_0t} f)(x)|^2$$

$$\leq |4\pi t|^{-n} \|f\|_1^2 \|V\|_2^2 ,$$

which converges to zero as $|t| \to \infty$.

(c) $\int_1^\infty \|V\, e^{-iK_o t} f\|\, dt \le \|f\|_1 \|V\|_2 \int_1^\infty (4\pi t)^{-\frac{n}{2}}\, dt < \infty$,

if $n \ge 3$. ∎

<u>COROLLARY 6.12</u> : Let $V \in L^p(\mathbb{R}^n)$, $f \in L^2(\mathbb{R}^n) \cap L^s(\mathbb{R}^n)$ with $2 \le p \le \infty$ and $s^{-1} = \frac{1}{2} + p^{-1}$. There is a constant α_n such that

$$\|V\, e^{-iK_o t} f\|_2 \le \alpha_n\, |t|^{-\frac{n}{p}}\, \|V\|_p\, \|f\|_s \ . \tag{6.26}$$

<u>Proof</u> : By the Hölder inequality (2.10)

$$\|V\, e^{-iK_o t} f\|_2 \le \|V\|_p\, \|e^{-iK_o t} f\|_q \ ,$$

with $q^{-1} = \frac{1}{2} - p^{-1}$. Let $g_t(x) = \exp(i\,\frac{x^2}{4t})\, f(x)$. Then, as in Lemma 6.10 :

$$\begin{aligned} (e^{-iK_o t} f)(x) &= (4\pi i t)^{-\frac{n}{2}}\, e^{i\frac{x^2}{4t}} \int e^{-i\frac{x.y}{2t}}\, e^{i\frac{y^2}{4t}}\, f(y)\, dy \\ &= (2it)^{-\frac{n}{2}}\, e^{i\frac{x^2}{4t}}\, \tilde{g}_t(\tfrac{x}{2t}) \ . \end{aligned}$$

Hence

$$\begin{aligned} \|e^{-iK_o t} f\|_q &= (4\pi|t|)^{-\frac{n}{2}}\, |2t|^{\frac{n}{q}} [\int dk\, |\tilde{g}_t(k)|^q]^{\frac{1}{q}} \\ &\le \alpha_n\, |t|^{\frac{n}{q}-\frac{n}{2}}\, \|g_t\|_s \ , \end{aligned}$$

where we have applied the Hausdorff-Young inequality (2.11), which is justified since $q^{-1} + s^{-1} = 1$ and $1 \le s \le 2$. Since $\|g_t\|_s = \|f\|_s$ and

$\frac{n}{q} - \frac{n}{2} = - \frac{n}{p}$ the result follows. ■

PROPOSITION 6.13 : Let $\mathcal{H} = L^2(\mathbb{R}^n)$, $n \geq 3$, $H_o = K_o$, $H = H_o + V$ with V real and $V \in L^p(\mathbb{R}^n)$, $2 \leq p < 3$ if $n = 3$ and $\frac{n}{2} < p < n$ if $n > 3$. Then $\Omega_\pm$ exist and $\sigma_{ac}(H) = [0,\infty)$.

Proof :

(i) The lower bound on p ensures that H is self-adjoint, see chapter 3. One has $\mathcal{H}_{ac}(K_o) = \mathcal{H}$. By Lemma 6.8, it suffices to show that $\Omega_\pm f$ exist for all f in some fundamental set $\mathcal{D}_o$, e.g. for each f_a.

Let

$$W_t\, f_a = e^{iHt}\, e^{-iK_o t}\, f_a$$

$$= e^{iHt}\, e^{-iK_o t}\, (K_o+i)^{-1}\, (K_o+i)\, f_a \,.$$

[Notice $f_a \in D(K_o)$]. It follows that

$$\frac{d}{dt} W_t\, f_a = i\, e^{iHt}\, (H-K_o)(K_o+i)^{-1}\, e^{-iK_o t}\, (K_o+i)\, f_a$$

$$= i\, e^{iHt}\, V(K_o+i)^{-1}\, e^{-iK_o t}\, (K_o+i)\, f_a \;,$$

which is strongly continuous since $\{e^{iHt}\}$ and $\{e^{iK_o t}\}$ are strongly continuous (Proposition 4.6) and $V(K_o+i)^{-1} \in \mathcal{B}(\mathcal{H})$, since V is K_o-bounded, see the end of chapter 4.
Now

$$\int_1^\infty d\tau\, \|\frac{d}{d\tau} W_\tau\, f_a\| = \int_1^\infty d\tau\, \|V\, e^{-iK_o\tau}\, f_a\| \;,$$

which is finite since by Corollary 6.12,

$$\|V e^{-iK_o t} f_a\| \le \alpha_n |t|^{-\frac{n}{p}} \|V\|_p \|f_a\|_s ,$$

$s^{-1} = \frac{1}{2} + p^{-1}$, and $\frac{n}{p} > 1$. (For n = 3, p = 2, one may alternatively use Corollary 6.11 (c)). Hence $\Omega_+ f_a$ exists by Lemma 6.9.
Similarly one obtains the existence of $\Omega_- f_a$.

(ii) We have $\sigma_{ac}(K_o) = \sigma(K_o) = [0,\infty)$ [see Chapter 4]. Since $\Omega_\pm$ exist, $[0,\infty) \subseteq \sigma_{ac}(H)$ [see Proposition 6.7]. Since V is K_o-compact, $\sigma_e(H) = \sigma_e(K_o)$ [Chapter 4]. Thus

$$[0,\infty) \subseteq \sigma_{ac}(H) \subseteq \sigma_e(H) = \sigma_e(K_o) = [0,\infty) .$$

Therefore

$$[0,\infty) = \sigma_{ac}(H) . \blacksquare$$

<u>Remark</u> : The proposition says that the essential spectrum of H contains an absolutely continuous part equal to $[0,\infty)$. There may also be eigenvalues or singularly continuous spectrum in $[0,\infty)$. This question will be studied in Chapters 9 and 10.

<u>Example</u> : Assume $|V(x)| \le (1 + |x|)^{-b}$. Then

$$\|V\|_p^p \le \int dx\ (1 + |x|)^{-bp} = c_n \int_0^\infty dx \frac{x^{n-1}}{(1+x)^{bp}} < \infty,$$

if $bp - n + 1 > 1$, i.e. $b > \frac{n}{p}$. By Proposition 6.13, $\Omega_\pm$ exist if $\frac{n}{p} > 1$.
Therefore $\Omega_\pm$ exist if $b > 1$.

This is the best possible result in the following sense : If $V(x) = \alpha|x|^{-b}$

with $0 < b \leq 1$, then it can be shown that $e^{iHt} e^{-iK_0 t}$ does not converge strongly (it converges weakly to zero, hence it cannot converge strongly to an isometric operator). For these and similar functions V, called *long range potentials*, $\Omega_\pm$ do not exist. In these cases one modifies the definition of the wave operators $\text{s-lim}_{t \to \pm\infty} e^{iHt} e^{-iK_0 t} T_t$, where T_t is a function of $P_k = -i \frac{\partial}{\partial x_k}$ $(k = 1,\ldots,n)$ and in general $T_{t+\tau} \neq T_t T_\tau$, see Chapter 12.

The last theorem shows that the existence of $\Omega_\pm$ is independent of the local behaviour of V(x) :

PROPOSITION 6.14 : Let $\mathcal{H} = L^2(\mathbb{R}^n)$, V real. Suppose that

$$\int_{|x|\geq R} |V(x)|^p \, dx < \infty \qquad (R < \infty) ,$$

with $2 \leq p < 3$ if $n = 3$, and $\frac{n}{2} < p < n$ if $n > 3$. Let H be a self-adjoint extension of $\hat{H} = K_0 + V$ with $D(\hat{H}) = \{f \in S(\mathbb{R}^n) \mid f(x) = 0 \text{ for } |x| < R\}$. Then $\Omega_\pm(H,K_0)$ exist.

Remark : $S(\mathbb{R}^n)$ is the set of C^∞-functions that decrease (together with all their derivatives) more rapidly than $|x|^{-m}$ for each $m = 1,2,\ldots$, as $|x| \to \infty$. $S(\mathbb{R}^n)$ is invariant under Fourier transformation. Thus $f \in S(\mathbb{R}^n)$ implies $\tilde{f} \in S(\mathbb{R}^n)$, hence $e^{-i|k|^2 t} |k|^2 \tilde{f}(k) \in S(\mathbb{R}^n)$. Therefore $e^{-iK_0 t} f \in D(K_0)$ for each t. Also $S(\mathbb{R}^n)$ is dense in $L^2(\mathbb{R}^n)$.

Proof : Clearly K_0 and V are defined on $D(\hat{H})$, and $\hat{H}^* \supseteq \hat{H}$.
Let $f \in S(\mathbb{R}^n)$ and $\phi \in C^\infty(\mathbb{R}^n)$ such that $0 \leq \phi(x) \leq 1$, $\phi(x) = 0$ for $|x| \leq R$, $\phi(x) = 1$ for $|x| > 2R$. Let Φ be the multiplication operator by $\phi(x)$. Then

$$e^{iHt} e^{-iK_0 t} f = e^{iHt} \Phi e^{-iK_0 t} f + e^{iHt} (I-\Phi) e^{-iK_0 t} f . \quad (6.27)$$

Since $(1-\phi) \in L^2(\mathbb{R}^n)$, one has

$$\|e^{iHt}(I-\Phi)e^{-iK_o t} f\| = \|(I-\Phi)e^{-iK_o t} f\| ,$$

which converges to zero as $t \to \pm\infty$, by Corollary 6.11(b).
Thus it suffices to show that $\operatorname{s-lim}_{t\to\pm\infty} e^{iHt}\,\Phi\, e^{-iK_o t} f$ exists. Now
$\Phi\, e^{-iK_o t} f \in D(\hat{H}) \subseteq D(H)$, and

$$\frac{d}{dt} e^{iHt}\,\Phi\, e^{-iK_o t} f = i\, e^{iHt}\,[\hat{H}\Phi - \Phi K_o]\, e^{-iK_o t} f$$

$$= i\, e^{iHt}\,[V\Phi + (K_o\Phi - \Phi K_o)]\, e^{-iK_o t} f.$$

$V\Phi$ is the multiplication operator by $V(x)\,\phi(x) \in L^p(\mathbb{R}^n)$, so that

$$\int_1^\infty dt\, \|e^{iHt}\, V\, \Phi\, e^{-iK_o t} f\| < \infty$$

as in Proposition 6.13. For $f \in \mathcal{S}(\mathbb{R}^n)$:

$$(K_o\, \Phi\, f)(x) = -\Delta(\phi f)(x)$$

$$= \phi(x)\,(-\Delta f(x)) - (\Delta\phi(x))\, f(x) - 2\sum_{j=1}^{n} \left(\frac{\partial\phi}{\partial x_j}\right)\left(\frac{\partial f}{\partial x_j}\right) . \quad (6.28)$$

Thus $K_o\, \Phi\, f - \Phi K_o f = -(\Delta\phi) f - 2i \sum_{j=1}^{n} \left(\frac{\partial\phi}{\partial x_j}\right) P_j\, f$, where
$(P_j f)(x) = -i\,\frac{\partial f}{\partial x_j}$ or $(\widetilde{P_j f})(k) = k_j\, \tilde{f}(k)$.

$(\Delta\phi)$ and $\left(\frac{\partial\phi}{\partial x_j}\right)$ are multiplication operators by L^2-functions. Hence by
Corollary 6.11 (c), one has

$$\int_1^\infty d\tau \, \| e^{iH\tau} (\Delta\phi) \, e^{-iK_o\tau} f \| < \infty \, ,$$

$$\int_1^\infty d\tau \, \| e^{iH\tau} (\frac{\partial\phi}{\partial x_j}) \, P_j \, e^{-iK_o\tau} f \| < \infty \, ,$$

since P_j commutes with $e^{-iK_o\tau}$ and $P_j f \in S(\mathbb{R}^n)$. By Lemma 6.9 $\underset{t \to \pm\infty}{\text{s-lim}} \, e^{iHt} \, \Phi \, e^{-iK_o t} f$ exists. ∎

Rough summary

$\Omega_\pm$ exist for potentials V with arbitrary local singularities if $|V(x)| \to 0$ as $|x| \to \infty$ faster than $\frac{1}{|x|}$, but not for potentials that decrease more slowly than $\frac{1}{|x|}$.

7 Completeness of the wave operators

We have that $\Omega_\pm$ maps $\mathcal{H}_{ac}(H_o)$ isometrically onto a subspace of $\mathcal{H}_{ac}(H)$. One says that $\Omega_\pm$ are *complete* if their range is equal to $\mathcal{H}_{ac}(H)$, i.e. if $F_+ = F_- = E_{ac}(H)$. The absolutely continuous parts of H and H_o are then unitarily equivalent.

LEMMA 7.1 : $\Omega_\pm$ are complete if and only if the limits $\text{s-lim}_{t\to\pm\infty} e^{iH_ot} e^{-iHt} E_{ac}(H)$ exist.

Proof : If $\Omega_\pm$ are complete, then

$$\text{s-lim}_{t\to\pm\infty} e^{iH_ot} e^{-iHt} E_{ac}(H) = \text{s-lim}_{t\to\pm\infty} e^{iH_ot} e^{-iHt} F_\pm = \Omega_\pm^*$$

exist, see Proposition 6.1(c).

Conversely, let $f \in \mathcal{H}_{ac}(H)$, $g_\pm = \text{s-lim}_{t\to\pm\infty} e^{iH_ot} e^{-iHt} f$. Then

$$\|f - e^{iHt} e^{-iH_ot} g_\pm\| = \|e^{iH_ot} e^{-iHt} f - g_\pm\| \to 0 \quad \text{as} \quad t \to \pm\infty \quad .$$

Hence $f = \Omega_\pm g$. Therefore

$$(\text{range } \Omega_\pm) = \mathcal{H}_{ac}(H) \ . \ \blacksquare$$

To prove completeness of Ω_+, it would suffice to show that $\int_1^\infty dt \, \|V e^{-iHt} f\| < \infty$ for all f in some dense subset of $\mathcal{H}_{ac}(H)$. Formally the problem is the same as that of the existence of $\Omega_\pm$, the roles of H and H_o being interchanged. However, if $H = K_o + V$, there is no explicit form for $(e^{-itH} f)(x)$, and the existence proof of $\Omega_\pm$ was based on the expression of $(e^{-itK_o} f)(x)$ given in Lemma 6.10. We therefore need different methods to prove completeness.

Hypotheses for Chapters 7, 9 and 11.

$$H = H_o + V \quad \text{with} \quad D(H) = D(H_o) \subseteq D(V).$$

Factorization of V : V = AB = BA with B H-bounded. A,B,H and H_o are self-adjoint. We then have

$$(Vg, e^{-iH_o t} Vg) = (Bg,[A\, e^{-iH_o t} A]\, Bg) \quad \forall g \in D(H).$$

Remark : If A is unbounded, it will be understood that $\|A\, e^{-iH_o t} A\| < \infty$ or $\|A(H_o-z)^{-1} A\| < \infty$ means that $A\, e^{-iH_o t} A$ or $A(H_o-z)^{-1} A$ respectively make sense as densely defined bounded operators.

LEMMA 7.2 : Let $f \in D(H)$. Assume

$$t \mapsto h(t) \equiv \|A\, e^{-iH_o t} A\| \in L^1(\mathbb{R}) \tag{7.1}$$

and

$$t \mapsto g(t) \equiv \|B\, e^{-iHt} f\| \in L^2(\mathbb{R}) . \tag{7.2}$$

Then $\text{s-}\lim_{t\to\pm\infty} e^{iH_o t} e^{-iHt} f$ exist.

Proof : We have

$$e^{iH_o t} e^{-iHt} f - e^{iH_o s} e^{-iHs} f = -i \int_s^t d\tau\, e^{iH_o \tau}\, V\, e^{-iH\tau} f.$$

By the Schwarz inequality, we get

$$\| e^{iH_o t} e^{-iHt} f - e^{iH_o s} e^{-iHs} f \|^2$$

$$= \int_s^t d\tau \int_s^t d\rho (e^{iH_o \tau} AB\, e^{-iH\tau} f, e^{iH_o \rho} AB\, e^{-iH\rho} f)$$

$$\leq \int_s^t d\tau \int_s^t d\rho \, \| Ae^{iH_o(\tau-\rho)} A \| \, \| Be^{-iH\tau} f \| \, \| Be^{-iH\rho} f \|$$

$$= \int_s^t d\tau \int_s^t d\rho \;\; h(\tau-\rho) \; g(\tau) \; g(\rho) = I_1 \, . \tag{7.3}$$

Substituting $\tau - x$ for ρ, and using again the Schwarz inequality, we obtain

$$I_1 \leq \int_{-\infty}^{+\infty} h(x) \; dx \int_s^t g(\tau) \; g(\tau-x) \; d\tau$$

$$\leq \int_{-\infty}^{+\infty} h(x) \; dx \; \{ \int_s^t d\tau \; |g(\tau)|^2 \}^{\frac{1}{2}} \{ \int_s^t d\tau \; |g(\tau-x)|^2 \}^{\frac{1}{2}}$$

$$\leq \| h \|_1 \; \{ \int_s^t d\tau \; |g(\tau)|^2 \}^{\frac{1}{2}} \; \| g \|_2 \; .$$

Since $h \in L^1$, $g \in L^2$, this converges to zero as $s,t \to +\infty$ or $s,t \to -\infty$. It follows that $e^{iH_o t} e^{-iHt} f$ is strongly Cauchy as $t \to \pm\infty$. ∎

<u>LEMMA 7.3</u> : Let $\Delta = (a,b]$ be a bounded interval, $B = B^*$ H-bounded and $C \in \mathcal{B}(\mathcal{H})$. Assume

$$\| \; B \, [(H-\lambda-i\varepsilon)^{-1} - (H-\lambda+i\varepsilon)^{-1}] C \| \; \leq M < \infty$$

for all $0 < \varepsilon \leq 1$ and all $\lambda \in (a,b + \delta_o)$, $\delta_o > 0$. Then

$\|B E_{\Delta_0} C\| \le (2\pi)^{-1} M |\Delta_0|$ for each subinterval $\Delta_0 = (c,d]$ of Δ and $\|B\phi(H) E_\Delta C\| \le (2\pi)^{-1} M \int_\Delta |\phi(\lambda)| d\lambda$ for each continuous function $\phi : \mathbb{R} \to \mathbb{C}$. ($|\Delta| \equiv b-a$).

Proof : Since B is H-bounded, $BE_{\Delta_0} = B(H+i)^{-1} (H+i) E_{\Delta_0}$ is in $\mathcal{B}(\mathcal{H})$. Let $f \in \mathcal{H}$, $g \in D(B)$. By Proposition 5.7 :

$$2\pi i(g,BE_{\Delta_0} Cf) = \lim_{\delta\to+0} \lim_{\varepsilon\to+0} \int_{c+\delta}^{d+\delta} d\lambda(Bg,[(H-\lambda-i\varepsilon)^{-1} - (H-\lambda+i\varepsilon)^{-1}] Cf)$$

$$= \lim_{\delta\to+0} \lim_{\varepsilon\to+0} \int_{c+\delta}^{d+\delta} d\lambda(g,B[(H-\lambda-i\varepsilon)^{-1} - (H-\lambda+i\varepsilon)^{-1}]Cf) .$$

Now $\|BE_{\Delta_0} C\| = \sup_{\|f\|=1} \sup_{\substack{g\in D(B)\\ \|g\|=1}} |(g,BE_{\Delta_0} Cf)|$

and

$$|(g,BE_{\Delta_0} Cf)| \le \sup_{0<\delta<\delta_0} \sup_{0<\varepsilon\le 1} (2\pi)^{-1} \int_{c+\delta}^{d+\delta} d\lambda \|g\| \|B[(H-\lambda-i\varepsilon)^{-1} - (H-\lambda+i\varepsilon)^{-1}]C\| \|f\|$$

$$\le (2\pi)^{-1} (d-c) \|g\| M\|f\|.$$

Thus

$$\|BE_{\Delta_0} C\| \le (2\pi)^{-1} M(d-c) . \tag{7.4}$$

(ii) Let $\Pi = \{\lambda_k ; \mu_k\}_{k=1}^n$ be a partition of Δ. Then

$$\|B\ \phi(H)\ E_\Delta C\| \le \left\|\int_\Delta \phi(\lambda)\ B\ dE_\lambda C\right\|$$

$$\le \lim_{|\Pi|\to 0} \sum_{k=1}^{n} \|\phi(\mu_k)\ B\ E_{(\lambda_{k-1},\lambda_k]} C\| . \qquad (7.5)$$

Then, by using (i) in the last inequality (7.5) one has

$$\lim_{|\Pi|\to 0} \sum_{k=1}^{n} \|\phi(\mu_k)\ B\ E_{(\lambda_{k-1},\lambda_k]} C\|$$

$$\le \lim_{|\Pi|\to 0} \sum_{k=1}^{n} |\phi(\mu_k)|\ M(2\pi)^{-1}\ |\lambda_k - \lambda_{k-1}|$$

$$= M(2\pi)^{-1} \int_a^b |\phi(\lambda)|\ d\lambda . \quad ∎$$

<u>LEMMA 7.4</u> : Under the hypotheses of Lemma 7.3, $t \mapsto \|B\ e^{-iHt}\ E_\Delta Cf\| \in L^2(\mathbb{R})$ for each $f \in \mathcal{H}$.

<u>Proof</u> : Let $0 < \varepsilon \le 1$. Then

$$I_\varepsilon \equiv \int_0^\infty \|B\ e^{-iHt}\ E_\Delta Cf\|^2\ e^{-\varepsilon t}\ dt + \int_{-\infty}^0 \|B\ e^{-iHt}\ E_\Delta Cf\|\ e^{\varepsilon t}\ dt$$

$$= \int_0^\infty (Cf, e^{iHt}(BE_\Delta)^*\ BE_\Delta\ e^{-iHt}\ Cf)\ e^{-\varepsilon t}\ dt \qquad (7.6)$$

$$+ \int_{-\infty}^0 (Cf, e^{iHt}(BE_\Delta)^*\ BE_\Delta\ e^{-iHt}\ Cf)\ e^{\varepsilon t}\ dt .$$

Then, by using Proposition 5.6 in (7.6), we get

$$I_\varepsilon = i \int_{\mathbb{R}} (Cf, [(H-\lambda+i\varepsilon)^{-1} - (H-\lambda-i\varepsilon)^{-1}]\ (BE_\Delta)^*\ BE_\Delta\ dE_\lambda\ Cf) \quad .$$

Let $\Pi = \{\lambda_k \ ; \ \mu_k\}$ be a partition of Δ. Then

$$I_\varepsilon \le \sup_{\Pi} \sum_{k=1}^{n} \|BE_\Delta[(H-\mu_k-i\varepsilon)^{-1} - (H-\mu_k+i\varepsilon)^{-1}]C\| \ \|f\|^2 \ \|BE_{(\lambda_{k-1},\lambda_k]}C\| \ .$$

Now by Lemma 7.3

$$\|BE_\Delta[(H-\mu-i\varepsilon)^{-1} - (H-\mu+i\varepsilon)^{-1}]C\| \le \frac{M}{2\pi}\int_\Delta \frac{2\varepsilon}{(\lambda-\mu)^2+\varepsilon^2}\, d\lambda$$

$$< \frac{M}{\pi}\int_{\mathbb{R}} \frac{\varepsilon}{x^2+\varepsilon^2}\, dx = \frac{M}{\pi}\int_{\mathbb{R}} \frac{1}{1+y^2}\, dy = M \ , \tag{7.7}$$

$$\|BE_{(\lambda_{k-1},\lambda_k]}C\| \le \frac{M}{2\pi}(\lambda_k - \lambda_{k-1}) \ . \tag{7.8}$$

This implies

$$I_\varepsilon \le \sup_{\Pi} \sum_{k=1}^{n} \frac{M^2}{2\pi}(\lambda_k - \lambda_{k-1}) \ \|f\|^2 = \frac{M^2}{2\pi}(b-a)\ \|f\|^2 \equiv \alpha_0 \quad .$$

As $\varepsilon \to +0$, the integrands in (7.6) are monotone increasing and converge to $\|B\, e^{-iHt}\, E_\Delta Cf\|^2$. By the monotone convergence theorem, $\|B\, e^{-iHt}\, E_\Delta Cf\|^2 \in L^1(\mathbb{R})$ and $\int_{\mathbb{R}} \|B\, e^{-iHt}\, E_\Delta Cf\|^2\, dt \le \alpha_0$. ∎

<u>LEMMA 7.5</u> : Let $\{W_z\}$ be a holomorphic family of compact operators in $\mathbb{C}\backslash\mathbb{R}$. Suppose that $I + W_z$ is invertible for each $z \in \mathbb{C}\backslash\mathbb{R}$ and that $W_{\lambda\pm i0} \equiv \lim W_{\lambda\pm i\varepsilon}$ as $\varepsilon \to +0$ exist in operator norm, the convergence being uniform in λ on each compact subset of $\mathbb{R}$. Define

$$\Gamma_0^\pm = \{\lambda \in \mathbb{R} \mid (I + W_{\lambda\pm i0}) \text{ is not invertible}\},$$

$$\Gamma_0 = \Gamma_0^+ \cup \Gamma_0^- \quad .$$

(a) Let Δ be a closed interval such that $\Delta \cap \Gamma_0 = \phi$, and define $\Delta_\pm = \{z = \lambda \pm i\varepsilon \mid \lambda \in \Delta,\ 0 \le \varepsilon \le 1\}$. Then $(I + W_z)^{-1}$ is norm continuous, in particular bounded, on each of the rectangles $\Delta_\pm$.

(b) Γ_0 is a closed set of Lebesgue measure zero called the *exceptional set* associated with $\{W_z\}$.

Proof : (for Γ_0^+) .

(i) By Lemma 2.4, $W_{\lambda \pm io}$ are compact.

(ii) Fix $z \notin \Gamma_0$. Let T be a finite rank operator such that $\|W_z - T\| < 1$. Then $(I + W_z - T)^{-1} \in \mathcal{B}(\mathcal{H})$ by the Neumann series, Proposition 2.3. Set $Y = T(I + W_z - T)^{-1}$. Then

$$(I + W_z) = (I + Y)(I + W_z - T) . \tag{7.9}$$

The range of T is a subspace N of dimension $n < \infty$. Let F be the projection with range N. Then $FY = Y$, i.e. $F + YF$ maps N into N. Let g be such that $(I + Y)g = 0$. Therefore $g = -Yg \in N$, i.e. $(F + YF)g = 0$.

It follows that $I + Y$ is invertible if and only if $F + YF$ is invertible in N. In our case, set $f = (I + W_z - T)^{-1}g$. By (7.9) :

$$(I + W_z)f = (I + Y)g = 0 . \tag{7.10}$$

Since $I + W_z$ is invertible, $f = 0$. Hence $g = 0$, i.e. $I + Y$ is invertible and $F + YF$ is invertible in N. $F + YF$ is given by an $n \times n$ matrix whose inverse we denote by $(F + YF)_N^{-1}$.

Let $Z = (F + YF)_N^{-1}\ F(I - Y + YF) + (I - F)$. Clearly $Z \in \mathcal{B}(\mathcal{H})$. After some algebra one finds that $Z(I + Y) = (I + Y)Z = I$. Hence

$(I + Y)^{-1} = Z \in \mathcal{B}(\mathcal{H})$. By (7.9) :

$$(I + W_z)^{-1} = (I + W_z - T)^{-1} (I + Y)^{-1} \quad , \tag{7.11}$$

which is in $\mathcal{B}(\mathcal{H})$.

(iii) Let $z, \xi \in \Delta_+$. By (ii) : $(I + W_z)^{-1}$, $(I + W_\xi)^{-1}$ are in $\mathcal{B}(\mathcal{H})$, and one has

$$I + W_z = [I - (W_\xi - W_z)(I + W_\xi)^{-1}] (I + W_\xi) . \tag{7.12}$$

It follows that

$$(I + W_z)^{-1} - (I + W_\xi)^{-1} = (I + W_\xi)^{-1} \{[I - (W_\xi - W_z)(I + W_\xi)^{-1}]^{-1} - I\}. \tag{7.13}$$

Since $\|W_z - W_\xi\| \to 0$ as $z \to \xi$, $[I - (W_\xi - W_z)(I + W_\xi)^{-1}]^{-1}$ exists if $|z - \xi|$ is small enough by the Neumann series. We have by using (7.13) and Proposition 2.3 :

$$| \|(I + W_z)^{-1}\| - \|(I + W_\xi)^{-1}\| | \leq \|(I + W_z)^{-1} - (I + W_\xi)^{-1}\|$$

$$\leq \|(I + W_\xi)^{-1}\| \sum_{n=1}^{\infty} [\|W_z - W_\xi\| \, \|(I + W_\xi)^{-1}\|]^n \tag{7.14}$$

$$= \|(I + W_\xi)^{-1}\|^2 \, \|W_z - W_\xi\| \, [1 - \|W_z - W_\xi\| \, \|(I + W_\xi)^{-1}\|]^{-1}.$$

Thus $\lim_{z \to \xi} \|(I + W_z)^{-1} - (I + W_\xi)^{-1}\| = 0$, and $z \mapsto \|(I + W_z)^{-1}\|$ is a continuous function on Δ_+. This proves (a).

(iv) Let $\lambda \in \mathbb{R}$. Choose $0 < \delta < 1$ such that $\|W_z - W_{\lambda+io}\| < \frac{1}{2}$ for all z with $\text{Im } z \geq 0$ and $|z - \lambda| \leq \delta$, i.e. all $z \in C_\lambda^+$.
Let T be a finite rank operator and such that $\|W_{\lambda+io} - T\| < \frac{1}{2}$. Then

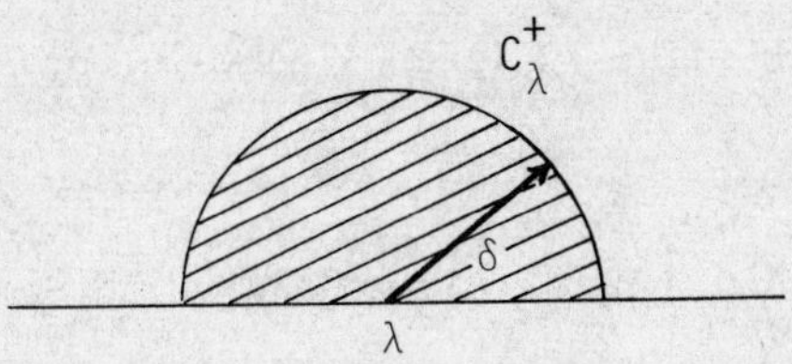

$$\|W_z - T\| \leq \|W_z - W_{\lambda+io}\| + \|W_{\lambda+io} - T\| < 1,$$

for all $z \in C_\lambda^+$.

Define F as in (ii) and set $Y_z = T(I + W_z - T)^{-1}$. One has :

(α) Y_z is holomorphic in the interior of C_λ^+ and continuous in norm up to the boundary [apply (iii) to $\{W_z - T\}$].

(β) $F + Y_zF$ is invertible in $N \iff I + W_z$ is invertible [by (ii)] $\iff z \notin \Gamma_0^+$. $F + Y_zF$ may be represented as an $n \times n$ matrix $\{a_{ik}(z)\}$. Let $\phi(z) = \det \{a_{ik}(z)\}$. Then

(α) ϕ is holomorphic in the interior of C_λ^+ and has continuous boundary values.

(β) $\phi(z) = 0 \iff z \in \Gamma_0^+$.

(γ) ϕ is not identically zero ($\phi(z) \neq 0$ if z is in the interior of C_λ^+).

An analytic function verifying (α), (β) and (γ) has the property that the set

$$\{z \mid z \in \partial C_\lambda^+ , \phi(z) = 0\}$$

is a closed set of (linear) Lebesgue measure zero [AS].

(v) By (iv), each $\lambda \in \mathbb{R}$ has a neighbourhood $N_\delta = [\lambda-\delta, \lambda+\delta]$ such that $\Gamma_0^+ \cap N_\delta$ is closed and of measure zero. Hence Γ_0^+ is closed and of measure zero, proving (b). ∎

Remark : Part (ii) of this proof is essentially the *Fredholm alternative* : If $A \in \mathcal{B}_\infty$ and $0 \neq z \in \mathbb{C}$, then either z is an eigenvalue of A or $(A-z)^{-1}$ is in $\mathcal{B}(\mathcal{H})$.

PROPOSITION 7.6 : Let H, H_0, A and B be self-adjoint, B H_0-bounded and H-bounded and $H = H_0 + AB$ with $D(H) = D(H_0)$. Assume that $\|A\, e^{-iH_0 t} A\| \in L^1(\mathbb{R})$, that $W_z \equiv B(H_0-z)^{-1}A$ satisfies the hypotheses of Lemma 7.5, that there exists $C \in \mathcal{B}(\mathcal{H})$ with dense range such that $\|B\, R^0_{\lambda+i\varepsilon} C\| \leq M < \infty$ for all $\lambda \in \mathbb{R}$ and $0 < \varepsilon \leq 1$, and that $\Omega_\pm$ exist. Then $\Omega_\pm$ are complete.

Proof : Let $\{E_\lambda\}$ be the spectral family of H and Γ_0 the exceptional set associated with $\{B(H_0-z)^{-1} A\}$ by Lemma 7.5.

(i) Since $|\Gamma_0| = 0$, one has $E_{\Gamma_0} \mathcal{H} \perp \mathcal{H}_{ac}(H)$. Then

$$\mathcal{H}_{ac}(H) \subseteq E_{\mathbb{R}\setminus\Gamma_0} \mathcal{H} . \tag{7.15}$$

We shall show

$$E_{\mathbb{R}\setminus\Gamma_0} \mathcal{H} \subseteq \Omega_\pm \mathcal{H} \quad . \tag{7.16}$$

By Proposition 6.2(a)

$$\Omega_\pm \mathcal{H} \subseteq \mathcal{H}_{ac}(H) \quad . \tag{7.17}$$

Using (7.15), (7.16) and (7.17) we obtain

$$\mathcal{H}_{ac}(H) = E_{\mathbb{R}\setminus\Gamma_0} \mathcal{H} = \Omega_\pm \mathcal{H} \quad . \tag{7.18}$$

It follows that

$$E_{\Gamma_0}\ \mathcal{H} = \mathcal{H}_s(H)\ .$$

Therefore

$$\sigma_s(H) = \sigma(H\ E_s(H)) \subseteq \Gamma_0\ . \tag{7.19}$$

Therefore the singular spectrum of H is a closed subset of Γ_0.

(ii) Proof of (7.16) : $\mathbb{R}\backslash\Gamma_0$ is open, hence $\mathbb{R}\backslash\Gamma_0 = \bigcup_{k=1}^{\infty} \Delta_k$ where $\{\Delta_k\}$ are disjoint open intervals. Therefore $E_{\mathbb{R}\backslash\Gamma_0} = \sum_{k=1}^{\infty} E_{\Delta_k}$, and it suffices to show that

$$E_{\Delta_k}\ \mathcal{H} \subseteq \Omega_{\pm}\ \mathcal{H} \quad \text{for each } k\ .$$

Let $\Delta_k = (a,b)$. Choose an increasing sequence of half-open intervals $(a_i,b_i] \subset (a,b)$, $a < a_i < b_i < b$ with $\lim_{i\to\infty} a_i = a$, $\lim_{i\to\infty} b_i = b$. Then

$$E_{(a,b)} = \text{s-}\lim_{i\to\infty} E_{(a_i,b_i]}\ .$$

Then

$$\bigcup_i E_{(a_i,b_i]}\ \mathcal{H} \quad \text{is dense in } E_{(a,b)}\ \mathcal{H}.$$

Therefore it suffices to show that $E_{(a_i,b_i]}\ \mathcal{H} \subseteq \Omega_{\pm}\ \mathcal{H}$ for each i. (7.20)

By Lemma 7.5,

$$\|(I + B(H_0-\lambda\mp i\varepsilon)^{-1}A)^{-1}\| \leq M_{(a_{i+1},b_{i+1})} < \infty$$

for all $\lambda \in (a_{i+1}, b_{i+1})$ and all $0 \leq \varepsilon \leq 1$.

By the second resolvent equation we have

$$B(H-z)^{-1}C = B(H_0-z)^{-1}C - B(H_0-z)^{-1} AB(H-z)^{-1}C.$$

Thus

$$B(H-z)^{-1}C = [I + B(H_0-z)^{-1}A]\, B(H_0-z)^{-1}C\ .$$

Hence

$$\|B(H-\lambda\mp i\varepsilon)^{-1}C\| \leq M_{(a_{i+1}, b_{i+1})} M$$

for all $\lambda \in (a_{i+1}, b_{i+1})$ and $0 \leq \varepsilon \leq 1$.

Therefore

$$\|B\, e^{-iHt}\, E_{(a_i,b_i]}Cf\| \in L^2(\mathbb{R}) \quad \text{for each } f \in \mathcal{H},$$

by Lemma 7.4. Since by hypothesis $\|AU_tA\| \in L^1(\mathbb{R})$, $\operatorname{s-lim}_{t\to\pm\infty} e^{-iH_0t}\, e^{iHt}\, E_{(a_i,b_i]}\, Cf$ exist by Lemma 7.2. It follows that $E_{(a_i,b_i]}Cf \in (\text{Range } \Omega_\pm)$ for each $f \in \mathcal{H}$. By hypothesis, $\{E_{(a_i,b_i]}Cf \mid f \in \mathcal{H}\}$ is dense in $E_{(a_i,b_i]}\mathcal{H}$. Therefore

$$E_{(a_i,b_i]}\, \mathcal{H} \subseteq \Omega_\pm\, \mathcal{H}\quad, \text{ proving (7.20)}\ . \quad \blacksquare$$

<u>LEMMA 7.7</u> : Let $n \geq 3$, and let $\phi, \psi : \mathbb{R}^n \to \mathbb{C}$ belong to $L^{p_1}(\mathbb{R}^n) \cap L^{p_2}(\mathbb{R})$ for some $2 \leq p_1 < n < p_2 < \infty$. Let S,T be the multiplication operators in $L^2(\mathbb{R}^n)$

by ϕ,ψ respectively. Then

(a) $\|S\, e^{-iK_0 t}\, T\| \in L^1(\mathbb{R})$ as a function of t.

(b) $\|S(K_0-\lambda\pm i\varepsilon)^{-1}\, T\| \leq M < \infty$ for all $\lambda \in \mathbb{R}$ and $\varepsilon > 0$.

(c) $\underset{\varepsilon\to+0}{\text{u-lim}}\, S(K_0-\lambda\mp i\varepsilon)^{-1}\, T$ exists for each $\lambda \in \mathbb{R}$ and the convergence is uniform in $\lambda \in \mathbb{R}$.

Proof :

(a) By Lemma 6.12, we have for i = 1,2 :

$$\|S\, e^{-iK_0 t}\, Tf\| \leq \alpha_n\, |t|^{-\frac{n}{p_i}}\, \|\phi\|_{p_i}\, \|Tf\|_{s_i}$$

with $s_i^{-1} = \frac{1}{2} + p_i^{-1}$. By applying Hölder's inequality one obtains

$$\|S\, e^{-iK_0 t}\, Tf\| \leq \alpha_n\, |t|^{-\frac{n}{p_i}}\, \|\phi\|_{p_i}\, \|\psi\|_{p_i}\, \|f\|_2 ,$$

i.e.
$$\|S\, e^{-iK_0 t}\, T\| \leq \alpha_n\, |t|^{-\frac{n}{p_i}}\, \|\phi\|_{p_i}\, \|\psi\|_{p_i} \qquad (i = 1,2) .$$

Since $\frac{n}{p_2} < 1$, $\|S\, e^{-iK_0 t}\, T\|$ is integrable over $[-1,+1]$.

Since $\frac{n}{p_1} > 1$, it is also integrable over $(-\infty,-1]$ and $[1,\infty)$.

(b) Let $\varepsilon \geq 0$ and set $Z_{\lambda+i\varepsilon} = i\int_0^\infty dt\, e^{i\lambda t}\, e^{-\varepsilon t}\, S\, e^{-iK_0 t}\, T$. By Proposition 5.2, we have $Z_{\lambda+i\varepsilon} \in \mathcal{B}(\mathcal{H})$, since

$$\|Z_{\lambda+i\varepsilon}\| \leq \int_0^\infty dt\, \|S\, e^{-iK_0 t}\, T\| \equiv M < \infty. \tag{7.21}$$

Now assume $f \in D(S^*)$ and $g \in D(T)$. We then have by Proposition 5.3, for $\varepsilon > 0$

$$i \int_0^\infty dt\, e^{i\lambda t} e^{-\varepsilon t} (f, S e^{-iK_0 t} Tg) = (S^* f, (K_0-\lambda-i\varepsilon)^{-1} Tg)$$

$$= (f, S(K_0-\lambda-i\varepsilon)^{-1} Tg) ,$$

since $S(K_0-\lambda-i\varepsilon)^{-1} \in \mathcal{B}(\mathcal{H})$ by Proposition 3.8 and Corollary 3.7. Hence $S(K_0-\lambda-i\varepsilon)^{-1} T = Z_{\lambda+i\varepsilon}$, and the assertion of (b) follows from (7.21).

(c) $$\| S(K_0-\lambda-i\varepsilon)^{-1} T - Z_{\lambda+io} \| \leq \int_0^\infty dt\, |e^{-\varepsilon t} - 1| \;\; \| S e^{-iK_0 t} T \| ,$$

which converges to zero as $\varepsilon \to 0$ by the Lebesgue dominated convergence theorem. Since the r.h.s. of the above inequality does not contain λ, the convergence is uniform in $\lambda \in \mathbb{R}$.
Similarly one finds

$$\underset{\varepsilon \to 0}{\text{u-lim}}\; S(K_0-\lambda+i\varepsilon)^{-1} T = -i \int_{-\infty}^0 dt\, e^{i\lambda t} S e^{-iK_0 t} T . \;\blacksquare$$

<u>Remark</u> : The proof shows that if ϕ and ψ are unbounded functions, the bounded operator $S e^{-iK_0 t} T$ can be naturally defined via some intermediate functions spaces. It may also be defined directly in $L^2(\mathbb{R}^n)$ as follows : Approximate ϕ and ψ in $L^{p_1} \cap L^{p_2}$-norm by a sequence of functions $\{\phi_m\}$ and $\{\psi_m\}$ respectively belonging to $L^{p_1} \cap L^{p_2} \cap L^\infty$, and define $S e^{-iK_0 t} T$ as the uniform limit of $S_m e^{-iK_0 t} T_m$.

<u>PROPOSITION 7.8</u> : Let $n \geq 3$, $\mathcal{H} = L^2(\mathbb{R}^n)$, $H = K_0 + V$ such that $V(x) = W_1(x)W_2(x) \in \mathbb{R}$ with $W_1(.) \in L^p(\mathbb{R}^n) \cap L^q(\mathbb{R}^n)$ for some $2 \leq p < n < q \leq \infty$ and $|W_2(x)| \leq (1 + |x|)^{-\alpha}$ for some $\alpha > 1$. Then $\Omega_\pm$ exist and are complete.

Remark that $W_1 \in L^{n-\varepsilon}(\mathbb{R}^n) \cap L^{n+\varepsilon}(\mathbb{R}^n)$ if $W_1 \in L^{n+\varepsilon}_{Loc}(\mathbb{R}^n)$ and $|W_1(x)| \leq C|x|^{-1-\varepsilon}$ for $|x|$ large.

Proof :

(i) Since $V(.) \in L^{n-\varepsilon}(\mathbb{R}^n)$, $\Omega_\pm$ exist by Proposition 6.13.

(ii) Let A be multiplication operator by $W_2(x)$, B the multiplication operator by $W_1(x)$. By Proposition 3.8 and Corollary 3.7, B is K_0-bounded with K_0-bound less than 1. Hence V is K_0-bounded and H-bounded (Proposition 3.6). It follows that B is H-bounded : for $f \in D(H)$ we have

$$\begin{aligned}
\|Bf\| &\leq \beta\|K_0 f\| + \alpha\|f\| \\
&\leq \beta\|Hf\| + \beta\|Vf\| + \alpha\|f\| \\
&\leq \beta\|Hf\| + \beta\beta'\|Hf\| + \beta\alpha'\|f\| + \alpha\|f\| \\
&\equiv \beta_0\|Hf\| + \alpha_0\|f\|.
\end{aligned}$$

We have also $\|A e^{-iK_0 t} A\| \in L^1(\mathbb{R})$ and $\|B e^{-iK_0 t} A\| \in L^1(\mathbb{R})$, by Lemma 7.7.

Completeness of $\Omega_\pm$ now follows from Proposition 7.6 if we set C = A and $W_z = B(K_0-z)^{-1}A$: Clearly C has dense range, W_z is compact by Proposition 2.8 and has boundary values as z approaches the real axis (Lemma 7.7). It only remains to show that $I + W_z$ is invertible if $\operatorname{Im} z \neq 0$, which will be done in (iii) below.

(iii) Let $\operatorname{Im} z \neq 0$ and set $R_z = (H-z)^{-1}$, $R_z^0 = (K_0-z)^{-1}$. Let $f \in \mathcal{H}$, $g \in D(B)$. We then have by the second resolvent equation (4.8) :

$$(f,(I + BR_z^0 A)(I - BR_z A)g)$$

$$= (f,g) - (Bf,[R_z - R_z^0 + R_z^0 V R_z]Ag) = (f,g)$$

and

$$(f,(I - BR_zA)(I + BR_z^oA)g)$$

$$= (f,g) - (Bf,[R_z - R_z^o + R_zVR_z^o]Ag) = (f,g) .$$

Hence $I + BR_z^oA$ is invertible and

$$(I + BR_z^oA)^{-1} = I - BR_zA \in \mathcal{B}(\mathcal{H}) . \blacksquare \tag{7.22}$$

Remark :

(a) The above proof can be extended to the case where W_2 satisfies the same hypotheses as W_1, in which case $V(.) \in L^{p_1}(\mathbb{R}^n) \cap L^{p_2}(\mathbb{R}^n)$ with $1 \leq p_1 < \frac{n}{2} < p_2 \leq \infty$.

(b) More will be said about asymptotic completeness in the next chapter. We shall in particular consider potentials decreasing more slowly to zero at infinity than those satisfying the assumption of Proposition 7.8.

(c) A different proof of asymptotic completeness, in the framework of time-dependent scattering theory, will be worked out in Chapter 12.

8 Restriction to spheres

For our subsequent developments of the theory it will be important to give a meaning to the restriction of the Fourier transform of certain functions in $L^2(\mathbb{R}^n)$ to spheres in $\mathbb{R}^n$. We shall do this in the language of operators by defining certain operators mapping $L^2(\mathbb{R}^n)$ into $L^2(S^{n-1})$, the set of all square integrable functions from the sphere S^{n-1} to $\mathbb{C}$ with respect to Lebesgue measure $d\omega$ on S^{n-1}.

We recall from chapter 4 that the spectral representation of K_o in $L^2(\mathbb{R}^n)$ is as follows :

$$\mathcal{U}_o : L^2(\mathbb{R}^n) \to L^2([0,\infty) , L^2(S^{n-1}))$$

with

$$(\mathcal{U}_o f)_\lambda (\omega) = 2^{-\frac{1}{2}} \lambda^{\frac{n-2}{4}} \tilde{f}(\sqrt{\lambda}\, \omega) .$$

Now if $f \in L^1(\mathbb{R}^n)$, then its Fourier transform is (equivalent to) a bounded and continuous function $\tilde{f}$. Thus the restriction of $\tilde{f}(k)$ to any sphere $|k|^2 = \lambda$ is well defined ; in other words $(\mathcal{U}_o f)_\lambda$ is defined for all λ and not merely almost everywhere.

Consider now the operator of multiplication in $L^2(\mathbb{R}^n)$ by a function ϕ with $\phi \in L^p(\mathbb{R}^n)$ for some $p \in [1,\infty]$. Then, by Hölder inequality, $\phi f \in L^1(\mathbb{R}^n)$ for each $f \in \mathcal{S}(\mathbb{R}^n)$, the Schwartz space of rapidly decreasing functions. We may therefore define, for each $\lambda > 0$ an operator $\hat{M}_\phi(\lambda) : \mathcal{S}(\mathbb{R}^n) \to L^2(S^{n-1})$ by

$$\hat{M}_\phi(\lambda) f = (\mathcal{U}_o \phi f)_\lambda. \tag{8.1}$$

The aim of this chapter is to show that, for certain ϕ, $\hat{M}_\phi(\lambda)$ can be extended to a bounded operator $M_\phi(\lambda)$ from $L^2(\mathbb{R}^n)$ to $L^2(S^{n-1})$, and to study the properties of $M_\phi(\lambda)$. (Notice that $\mathcal{S}(\mathbb{R}^n)$ is dense in $L^2(\mathbb{R}^n)$, so that in all cases $\hat{M}_\phi(\lambda)$ is densely defined). We shall always use the notation $M_\phi(\lambda)$ for the closure of $\hat{M}_\phi(\lambda)$.

PROPOSITION 8.1 : Assume $n \geq 2$ and $\phi \in L^2(\mathbb{R}^n)$. Then, for each $\lambda > 0$, $M_\phi(\lambda)$ is a Hilbert-Schmidt operator from $L^2(\mathbb{R}^n)$ to $L^2(S^{n-1})$, and the function $\lambda \mapsto M_\phi(\lambda)$ is continuous in Hilbert-Schmidt norm.

Proof : $\hat{M}_\phi(\lambda)$ is an integral operator with kernel

$$a(\omega,x) = 2^{-\frac{1}{2}} \lambda^{\frac{n-2}{4}} (2\pi)^{-\frac{n}{2}} e^{-i\sqrt{\lambda}\omega.x} \phi(x) . \tag{8.2}$$

By the two-Hilbert space version of Proposition 2.6,

$$\|M_\phi(\lambda)\|^2_{HS} = \int_{S^{n-1}} d\omega \int_{\mathbb{R}^n} dx \, |a(\omega,x)|^2$$

$$= \frac{1}{2} (2\pi)^{-n} \lambda^{\frac{n}{2}-1} \|\phi\|_2^2 \, |S_{n-1}| < \infty,$$

so that $M_\phi(\lambda)$ is Hilbert-Schmidt. Similarly

$$\|M_\phi(\lambda) - M_\phi(\mu)\|^2_{HS} = \frac{1}{2} (2\pi)^{-n} \iint d\omega \, dx \, |\lambda^{\frac{n-2}{4}} e^{-i\sqrt{\lambda}\omega.x}$$

$$- \mu^{\frac{n-2}{4}} e^{-i\sqrt{\mu}\omega.x}|^2 \, |\phi(x)|^2 ,$$

which converges to zero as $\mu \to \lambda$ by Lebesgue dominated convergence theorem.

■

PROPOSITION 8.2 : Assume $n \geq 3$ and $\phi \in L^{p_1}(\mathbb{R}^n) \cap L^{p_2}(\mathbb{R}^n)$ with $2 \leq p_1 < n < p_2 \leq \infty$. Then, for each $\lambda > 0$, $M_\phi(\lambda)$ is a compact operator from $L^2(\mathbb{R}^n)$ to $L^2(S^{n-1})$, and the function $\lambda \mapsto M_\phi(\lambda)$ is continuous in the operator norm.

Proof :

(i) We have for $f \in \mathcal{S}(\mathbb{R}^n)$

$$(\phi f, [(K_0-\lambda-i\varepsilon)^{-1} - (K_0-\lambda+i\varepsilon)^{-1}]\, \phi f)$$

$$= \int_0^\infty dk\, k^{n-1} \int_{S^{n-1}} d\omega\, |(\widetilde{\phi f})(k\omega)|^2 \left(\frac{1}{k^2-\lambda-i\varepsilon} - \frac{1}{k^2-\lambda+i\varepsilon}\right)$$

$$= \frac{1}{2}\int_0^\infty \mu^{\frac{n-2}{2}} d\mu \int_{S^{n-1}} d\omega\, |(\widetilde{\phi f})(\sqrt{\mu}\omega)|^2 \frac{2i\varepsilon}{(\mu-\lambda)^2 + \varepsilon^2} \, . \qquad (8.3)$$

Since $\phi f \in L^1(\mathbb{R}^n) \cap L^2(\mathbb{R}^n)$, the function $\mu \mapsto \mu^{\frac{n-2}{2}} \int |(\widetilde{\phi f})(\sqrt{\mu}\omega)|^2 d\omega$ is continuous and in $L^1((0,\infty))$. Hence, as $\varepsilon \to 0$, the r.h.s. of (8.3) converges to

$$2\pi i\, \frac{1}{2}\, \lambda^{\frac{n-2}{2}} \int d\omega\, |(\widetilde{\phi f})(\sqrt{\lambda}\omega)|^2 = 2\pi i\, \|\hat{M}_\phi(\lambda) f\|^2 \, .$$

Let S denote the multiplication operator by $\phi(x)$. Then, by Lemma 7.7, $Z_{\lambda\pm io} \equiv \lim_{\varepsilon\to 0} S^*(K_0-\lambda\mp i\varepsilon)^{-1} S \in \mathcal{B}(\mathcal{H})$ and (8.3) leads to

$$\|\hat{M}_\phi(\lambda) f\|^2 = \frac{1}{2\pi i} (f, [Z_{\lambda+io} - Z_{\lambda-io}] f)$$

$$\leq \frac{1}{2\pi} [\|Z_{\lambda+io}\| + \|Z_{\lambda-io}\|]\, \|f\|^2$$

$$\leq c(\|\phi\|_{p_1}^2 + \|\phi\|_{p_2}^2)\, \|f\|^2 ,$$

(see the proof of Lemma 7.7 for the last inequality). Hence $\hat{M}_\phi(\lambda)$ is bounded on $S(\mathbb{R}^n)$, i.e. $M_\phi(\lambda)$ is a bounded operator with

$$\|M_\phi(\lambda)\| \le c_1(\|\phi\|_{p_1} + \|\phi\|_{p_2}) . \tag{8.4}$$

(ii) Let $\{\phi_i\}$ be a sequence of functions belonging to $L^2(\mathbb{R}^n) \cap L^{p_1}(\mathbb{R}^n) \cap L^{p_2}(\mathbb{R}^n)$ such that $\|\phi - \phi_i\|_{p_k} \to 0$ as $i \to \infty$, for $k = 1,2$. By (8.4),

$$\|M_\phi(\lambda) - M_{\phi_i}(\mu)\| \to 0 \qquad \text{as} \quad i \to \infty. \tag{8.5}$$

Since M_{ϕ_i} is Hilbert-Schmidt by Proposition 8.1, this shows that $M_\phi(\lambda)$ is compact.

(iii)
$$\|M_\phi(\lambda) - M_\phi(\mu)\| \le \|M_\phi(\lambda) - M_{\phi_i}(\lambda)\|$$
$$+ \|M_{\phi_i}(\lambda) - M_{\phi_i}(\mu)\| + \|M_{\phi_i}(\mu) - M_\phi(\mu)\|.$$

By using (8.5) for the first and the third term on the r.h.s. and Proposition 8.1 for the second term, one arrives at the continuity in norm of $M_\phi(\lambda)$. ∎

PROPOSITION 8.3 : Let $n \ge 3$ and let $\phi,\psi \in L^{p_1}(\mathbb{R}^n) \cap L^{p_2}(\mathbb{R}^n)$ with $2 \le p_1 < n < p_2 \le \infty$. Define $Z_{\lambda\pm io}$ by

$$Z_{\lambda\pm io} = \underset{\varepsilon\to+0}{\text{u-lim}}\ \phi(Q)^* \ (K_o - \lambda \mp i\varepsilon)^{-1}\ \psi(Q) .$$

Then

$$Z_{\lambda+io} - Z_{\lambda-io} = 2\pi i\ M_\phi(\lambda)^*\ M_\psi(\lambda) . \tag{8.6}$$

Proof : As in (i) of the proof of Proposition 8.2, one obtains for $f,g \in \mathcal{S}(\mathbb{R}^n)$ that

$$(f,(Z_{\lambda+io} - Z_{\lambda-io})g) = \lim_{\varepsilon \to +0} (\phi f,[(K_o-\lambda-i\varepsilon)^{-1} - (K_o - \lambda+i\varepsilon)^{-1}]\psi g)$$

$$= 2\pi i \ (M_\phi(\lambda)f,M_\psi(\lambda)g) = 2\pi i \ (f,M_\phi(\lambda)^* \ M_\psi(\lambda)g). \ \blacksquare$$

PROPOSITION 8.4 : Assume the hypotheses of Proposition 8.2. Let $g \in \mathcal{S}(\mathbb{R}^n)$, let $\psi : \mathbb{R} \to \mathbb{C}$ be bounded and continuous, and $\Delta \subseteq \mathbb{R}$ an interval. Then

(a) $$\bar{\phi} \ E^o_\Delta g \equiv \phi(Q)^* \ E^o_\Delta g = \int_\Delta M_\phi(\lambda)^* \ (\mathcal{U}_o g)_\lambda \ d\lambda . \tag{8.7}$$

(b) $$\phi(Q)^* \ \psi(K_o)g = \int_0^\infty \psi(\lambda) \ M_\phi(\lambda)^* \ (\mathcal{U}_o g)_\lambda \ d\lambda . \tag{8.8}$$

(c) $$\operatorname*{s-lim}_{\varepsilon \to +0} \phi(Q)^* \ [(K_o-\lambda-i\varepsilon)^{-1} - (K_o-\lambda+i\varepsilon)^{-1}]g$$

$$= 2\pi i \ M_\phi(\lambda)^* \ (\mathcal{U}_o g)_\lambda. \tag{8.9}$$

(d) Let $h : \Delta \to L^2(\mathbb{R}^n)$ be strongly continuous and bounded. Then

$$\int_\Delta (h(\lambda),\phi(Q)^* \ dE^o_\lambda g) = \int_\Delta (h(\lambda),M_\phi(\lambda)^* \ (\mathcal{U}_o g)_\lambda) \ d\lambda . \tag{8.10}$$

Proof :

(i) By Proposition 8.2 and (8.4), $\lambda \mapsto M_\phi(\lambda)^*$ is bounded and continuous from $(0,\infty)$ to $\mathcal{B}(L^2(S^{n-1}), L^2(\mathbb{R}^n))$ (with respect to the operator norm). Since $g \in \mathcal{S}(\mathbb{R}^n)$, $\lambda \mapsto (\mathcal{U}_o g)_\lambda$ is strongly continuous from $(0,\infty)$ to $L^2(S^{n-1})$ and rapidly decreasing as $\lambda \to \infty$. Hence $\lambda \mapsto \psi(\lambda) \ M_\phi(\lambda)^* \ (\mathcal{U}_o g)_\lambda$ is strongly continuous from $(0,\infty)$ to $L^2(\mathbb{R}^n)$, and $\lambda \mapsto \|\psi(\lambda) \ M_\phi(\lambda)^* \ (\mathcal{U}_o g)_\lambda\|$ is in $L^1(0,\infty)$. By

Proposition 5.1(c), $\int_\Gamma \psi(\lambda)\ M_\phi(\lambda)^*\ (\mathcal{U}_0 g)_\lambda$ exists and defines a vector in $L^2(\mathbb{R}^n)$ for each interval $\Gamma \subseteq \mathbb{R}$.

(ii) Let now $f \in D(\phi(Q))$ and $\Gamma \subseteq \mathbb{R}$ an interval. Then [Problem 14.1]

$$(\psi(K_0)\ E_\Gamma^0 g, \phi(Q) f) = \int_\Gamma (\psi(\lambda)(\mathcal{U}_0 g)_\lambda, (\mathcal{U}_0 \phi f)_\lambda)\ d\lambda$$

$$= \int_\Gamma (\psi(\lambda)(\mathcal{U}_0 g)_\lambda, M_\phi(\lambda) f)\ d\lambda$$

$$= \int_\Gamma (\psi(\lambda)\ M_\phi(\lambda)^*\ (\mathcal{U}_0 g)_\lambda, f)\ d\lambda$$

$$= (\int_\Gamma \psi(\lambda)\ M_\phi(\lambda)^*\ (\mathcal{U}_0 g)_\lambda\ d\lambda,\ f)\ ,$$

since the integral defines a vector in $\mathcal{H}$. Hence $\psi(K_0)\ E_\Gamma^0 g \in D(\phi(Q)^*)$ and

$$\phi(Q)^*\ \psi(K_0)\ E_\Gamma^0 g = \int_\Gamma \psi(\lambda)\ M_\phi(\lambda)^*\ (\mathcal{U}_0 g)_\lambda\ d\lambda.$$

Taking $\Gamma = \Delta$ and $\psi \equiv 1$, we obtain (a), whereas (b) follows by setting $\Gamma = \mathbb{R}$.

(iii) Let $g' = (1 + |Q|)^n g$. We have $g' \in S(\mathbb{R}^n)$. From Lemma 7.7 we may deduce the existence of the following limit

$$\underset{\varepsilon \to +0}{\text{s-lim}}\ \phi(Q)^*\ [(K_0-\lambda-i\varepsilon)^{-1} - (K_0-\lambda+i\varepsilon)^{-1}]g \tag{8.11}$$

$$= \underset{\varepsilon \to +0}{\text{s-lim}}\ \phi(Q)^*\ [(K_0-\lambda-i\varepsilon)^{-1} - (K_0-\lambda+i\varepsilon)^{-1}]\ (1 + |Q|)^{-n} g'\ .$$

On the other hand, for $f \in S(\mathbb{R}^n)$, we obtain as in part (i) of the Proposition 8.2 that

$$\lim_{\varepsilon \to 0} (f,\phi(Q)^* [(K_0-\lambda-i\varepsilon)^{-1} - (K_0-\lambda+i\varepsilon)^{-1}]g)$$

$$= \lim_{\varepsilon \to 0} \frac{1}{2}\int_0^\infty d\mu\ \mu^{\frac{n-2}{2}} \int d\omega\ \overline{\phi\tilde{f}}(\sqrt{\mu}\omega)\tilde{g}(\sqrt{\mu}\omega)\ \frac{2i\varepsilon}{(\mu-\lambda)^2+\varepsilon^2}$$

$$= 2\pi i\ ((\mathcal{U}_0 f)_\lambda,\ (\mathcal{U}_0 g)_\lambda)$$

$$= 2\pi i\ (f,\ M_\phi(\lambda)^*\ (\mathcal{U}_0 g)_\lambda)\ .$$

Since $\mathcal{S}(\mathbb{R}^n)$ is dense in $L^2(\mathbb{R}^n)$, it follows that

$$\underset{\varepsilon \to +0}{\text{s-lim}}\ \phi(Q)^* [(K_0-\lambda-i\varepsilon)^{-1} - (K_0-\lambda+i\varepsilon)^{-1}]g$$

$$= 2\pi i\ M_\phi(\lambda)^*\ (\mathcal{U}_0 g)_\lambda\ ,$$

which proves (8.9).

(iv) It suffices to prove (d) for each bounded interval Δ. In that case, let $\Pi = \{\lambda_k\ ;\ \mu_k\}$ be a partition of Δ. Let $\Delta_k = (\lambda_{k-1},\lambda_k]$. One has

$$\int_\Delta (h(\lambda),\phi(Q)^*\ dE^0_\lambda g) = \lim_{|\Pi|\to 0} \sum_{k=1}^n (h(\mu_k),\phi(Q)^*\ E^o_{\Delta_k} g)$$

$$= \lim_{|\Pi|\to 0} \sum_{k=1}^n \int_{\Delta_k} (h(\mu_k),M_\phi(\lambda)^*\ (\mathcal{U}_0 g)_\lambda)\ d\lambda$$

$$= \lim_{|\Pi|\to 0} \sum_{k=1}^n \int_{\Delta_k} (h(\lambda),M_\phi(\lambda)^*\ (\mathcal{U}_0 g)_\lambda)\ d\lambda \qquad (8.12)$$

$$+ \lim_{|\Pi|\to 0} \sum_{k=1}^n \int_{\Delta_k} (h(\mu_k) - h(\lambda),M_\phi(\lambda)^*\ (\mathcal{U}_0 g)_\lambda)\ d\lambda$$

$$= \int_\Delta (h(\lambda),M_\phi(\lambda)^*\ (\mathcal{U}_0 g)_\lambda)\ d\lambda + \lim_{|\Pi|\to 0} R(\Pi)$$

if we suppose that $R(\Pi)$ is equal to the last term of (8.12). Let $\varepsilon > 0$. Since Δ is compact, there is $\delta > 0$ such that $\|h(\mu) - h(\lambda)\| < \varepsilon$ whenever $\mu,\lambda \in \Delta$ and $|\mu-\lambda| < \delta$. Thus, if $|\Pi| < \delta$:

$$|R(\Pi)| < \varepsilon \int_\Delta \|M_\phi(\lambda)^* (\mathcal{U}_0 g)_\lambda\| \, d\lambda ,$$

which shows that $R(\Pi) \to 0$ as $|\Pi| \to 0$. This proves (d). ∎

In Proposition 8.2 we have essentially shown the following result : Let S be an operator in $L^2(\mathbb{R}^n)$ such that $S.\mathcal{S}(\mathbb{R}^n) \subseteq L^1(\mathbb{R}^n) \cap L^2(\mathbb{R}^n)$ and such that $S^*(K_0-\lambda\mp i\varepsilon)^{-1} S$ converge weakly to a bounded operator as $\varepsilon \to 0$. Then the operator $\hat{M}_S(\lambda)$ defined by

$$\hat{M}_S(\lambda)f = (\mathcal{U}_0 Sf)_\lambda \quad , \quad f \in \mathcal{S}(\mathbb{R}^n) ,$$

is bounded.

We shall now obtain some kind of a converse of this : If $\hat{M}_S(\lambda)$ is bounded and Hölder continuous as a function of λ, then $S^*(K_0-\lambda\mp i\varepsilon)^{-1} S$ converge as $\varepsilon \to 0$. This will allow us to extend the completeness result of Proposition 7.8 to a larger class of functions V (as is usual in physics, the function V will henceforth be called a *potential*).

<u>PROPOSITION 8.5</u> : Let S,T be operators in $\mathcal{B}(L^2(\mathbb{R}^n))$ mapping $\mathcal{S}(\mathbb{R}^n)$ into $L^1(\mathbb{R}^n)$ and such that $\hat{M}_S(\lambda)$ and $\hat{M}_T(\lambda)$ are bounded for each $\lambda > 0$. Assume also that their closures $M_S(.)$ and $M_T(.)$ are locally Hölder continuous on $(0,\infty)$, in other words that for each bounded interval in $(0,\infty)$ there is a constant c_Δ and a number $\gamma > 0$ such that

$$\left.\begin{aligned} \|M_S(\lambda) - M_S(\mu)\| &\le c_\Delta\, |\lambda-\mu|^\gamma \\ \|M_T(\lambda) - M_T(\mu)\| &\le c_\Delta\, |\lambda-\mu|^\gamma . \end{aligned}\right\} \quad \forall\lambda,\mu \in \Delta.$$

Then

(a) for each $\lambda > 0$, the following limits exist :

$$\underset{\varepsilon \to +0}{\text{u-lim}}\ S^*(K_0-\lambda \mp i\varepsilon)^{-1}\ T \equiv Z_{\lambda \pm io}$$

and the convergence is uniform on each compact subset of $(0,\infty)$.

(b) $$Z_{\lambda+io} - Z_{\lambda-io} = 2\pi i\ M_S(\lambda)^*\ M_T(\lambda)\ .$$

<u>Proof</u> :

(i) Let $\lambda > 0$. Choose an interval $\Delta = [a,b]$ in $(0,\infty)$ containing λ in its interior, and set $\Delta' = \mathbb{R}\setminus\Delta$. Then

$$\begin{aligned} S^*(K_0-z)^{-1}\ T &= S^*(K_0-z)^{-1}\ E^o_\Delta\ T + S^*(K_o-z)^{-1}\ E^o_{\Delta'}\ T \\ &= S^*(K_o-z)^{-1}\ E^o_\Delta\ T + S^*(K_o\ E^o_{\Delta'}-z)^{-1}\ E^o_{\Delta'}\ T\ . \end{aligned} \tag{8.13}$$

The last term is holomorphic (with respect to the operator norm) in the domain $\{z \mid a < \text{Re } z < b\}$ in particular it is well defined and continuous at the point $z = \lambda$.

The first term on the r.h.s. of (8.13) may be written as in part (i) of the proof of Proposition 8.2, in the following way :

$$\begin{aligned} S^*(K_0-z)^{-1}\ E^o_\Delta\ T &= \int_\Delta (\mu-z)^{-1}\ M^*_S(\mu)\ M_T(\mu)\ d\mu \\ &= \int_\Delta (\mu-z)^{-1}\ [M^*_S(\mu)\ M_T(\mu) - M^*_S(\lambda)\ M_T(\lambda)]\ d\mu \\ &\quad + M^*_S(\lambda)\ M_T(\lambda)\ \text{Log}\ \frac{b-z}{a-z}\ . \end{aligned} \tag{8.14}$$

Setting $z = \lambda \pm i\varepsilon$, one sees that the last term converges as $\varepsilon \to 0$, since $\lambda \in (a,b)$. To deal with the first term on the r.h.s. of (8.14) we notice that

$$\int_{\Delta} (\mu-\lambda)^{-1} [M_S^*(\mu)\, M_T(\mu) - M_S^*(\lambda)\, M_T(\lambda)]\, d\mu \tag{8.15}$$

exists (e.g. as improper Riemann integral in the uniform operator topology), since $\mu \mapsto M_S^*(\mu)\, M_T(\mu)$ is Hölder continuous, i.e.

$$\| M_S^*(\mu)\, M_T(\mu) - M_S^*(\lambda)\, M_T(\lambda) \|$$

$$\leq c_{\Delta} |\lambda - \mu|^{\gamma} \sup_{\sigma \in \Delta} \{ \| M_S(\sigma) \| + \| M_T(\sigma) \| \}.$$

Furthermore, setting $A_{ST}(\mu,\lambda) = M_S^*(\mu)\, M_T(\mu) - M_S^*(\lambda)\, M_T(\lambda)$, one has

$$\begin{aligned} &\int_{\Delta} \left(\frac{1}{\mu-\lambda\mp i\varepsilon} - \frac{1}{\mu-\lambda}\right) A_{ST}(\mu,\lambda)\, d\mu \\ &= \pm i\varepsilon \int_{\Delta} \frac{1}{(\mu-\lambda\mp i\varepsilon)(\mu-\lambda)} A_{ST}(\mu,\lambda)\, d\mu \\ &= \pm i\varepsilon \int_{\lambda-\delta}^{\lambda+\delta} \frac{1}{(\mu-\lambda\mp i\varepsilon)(\mu-\lambda)} A_{ST}(\mu,\lambda)\, d\mu \\ &\quad \pm i\varepsilon \int_{a}^{\lambda-\delta} \frac{1}{(\mu-\lambda\mp i\varepsilon)(\mu-\lambda)} A_{ST}(\mu,\lambda)\, d\mu \\ &\quad \pm i\varepsilon \int_{\lambda+\delta}^{b} \frac{1}{(\mu-\lambda\mp i\varepsilon)(\mu-\lambda)} A_{ST}(\mu,\lambda)\, d\mu . \end{aligned} \tag{8.16}$$

Let now $\eta > 0$. We shall show that one may choose $\delta > 0$ and $\varepsilon_0 > 0$ such that the norm of the r.h.s. of (8.16) is less than η for all $0 < \varepsilon \leq \varepsilon_0$. This then shows that the limit of the first term on the r.h.s. of (8.14) is given by (8.15) and establishes the existence of u-lim $S^*(K_0-\lambda\mp i\varepsilon)^{-1}\, T$ as $\varepsilon \to 0$.

We now estimate the integrals on the r.h.s. of (8.16). In the first integral, we use $|\mu-\lambda\mp i\varepsilon|^{-1} \le \varepsilon^{-1}$ and obtain that

$$\varepsilon \int_{\lambda-\delta}^{\lambda+\delta} \left\| \frac{1}{(\mu-\lambda\mp i\varepsilon)(\mu-\lambda)} A_{ST}(\mu,\lambda) \right\| d\mu$$

$$\le \int_{\lambda-\delta}^{\lambda+\delta} \frac{1}{|\mu-\lambda|} \|A_{ST}(\mu,\lambda)\| \, d\mu < \frac{\eta}{2}$$

provided δ is sufficiently small. In the remaining two integrals, we use $|\mu - \lambda|^{-1} \le \delta^{-1}$ and $|\mu - \lambda\mp i\varepsilon|^{-1} \le \delta^{-1}$. Hence

$$\varepsilon \int_{a}^{\lambda-\delta} \left\| \frac{1}{(\mu-\lambda\mp i\varepsilon)(\mu-\lambda)} A_{ST}(\mu,\lambda) \right\| d\mu$$

$$+ \varepsilon \int_{\lambda+\delta}^{b} \left\| \frac{1}{(\mu-\lambda\mp i\varepsilon)(\mu-\lambda)} A_{ST}(\mu,\lambda) \right\| d\mu$$

$$\le \varepsilon \, \delta^{-2} \int_{a}^{b} \|A_{ST}(\mu,\lambda)\| \, d\mu < \frac{\eta}{2}$$

provided $\varepsilon \le \varepsilon_o$ for some suitable ε_o.

(ii) Let Γ be a compact subset of $(0,\infty)$, and set $\Gamma^{\pm} = \{z = \lambda\pm i\sigma | \lambda \in \Gamma, \ 0 \le \sigma \le 1\}$. Then, by (i), $S^*(K_o-z)^{-1}T$ is continuous (in operator norm) on Γ^+ and on Γ^-, hence uniformly continuous. This shows that the limits $Z_{\lambda\pm io}$ exist uniformly in $\lambda \in \Gamma$.

(iii) (b) is obtained in the same way as Proposition 8.3. ∎

We shall now show that Proposition 8.5 applies in the case where S and T are multiplication operators in $L^2(\mathbb{R}^n)$ by functions of the form $\phi(x) = (1 + |x|)^{-\beta}$, $\beta > \frac{1}{2}$. The proof requires a few preliminary results which we give in Lemmas 8.6 through 8.8.

LEMMA 8.6 : (Restriction to a hyperplane) : Let $\beta > \frac{1}{2}$, $\lambda \in \mathbb{R}$, and denote by $\hat{T}_\beta(\lambda)$ the following operator from $L^2(\mathbb{R}^n)$ to $L^2(\mathbb{R}^{n-1})$ with domain $D(\hat{T}_\beta(\lambda)) = S(\mathbb{R}^n)$:

$$(\hat{T}_\beta(\lambda)f)(\xi) = [F(1 + |Q|)^{-\beta} f] (\xi_1,\ldots,\xi_{n-1},\lambda) \qquad (8.17)$$

where $\xi = (\xi_1,\ldots,\xi_{n-1}) \in \mathbb{R}^{n-1}$. $\hat{T}_\beta(\lambda)$ is bounded and hence its closure $T_\beta(\lambda)$ belongs to $B(L^2(\mathbb{R}^n) , L^2(\mathbb{R}^{n-1}))$ (the set of all bounded, everywhere defined operators from $L^2(\mathbb{R}^n)$ to $L^2(\mathbb{R}^{n-1})$) . The function $\lambda \mapsto T_\beta(\lambda)$ is Hölder continuous in the operator norm for each Hölder exponent $\gamma < \min (\beta - \frac{1}{2}, 1)$.

Proof : For $x \in \mathbb{R}^n$, we write $\hat{x} = (x_1,\ldots,x_{n-1}) \in \mathbb{R}^{n-1}$. One has for $f \in S(\mathbb{R}^n)$:

$$|(F(1 + |Q|)^{-\beta}f)(\xi,\lambda)|$$

$$\leq (2\pi)^{-\frac{n}{2}} \int dx_n(1 + |x_n|)^{-\beta}|\int d\hat{x}\ e^{-i\xi.\hat{x}} \left[\frac{1 + |x_n|}{1 + |x|}\right]^\beta f(\hat{x},x_n)|$$

$$\leq (2\pi)^{-\frac{n}{2}} [\int dx_n(1 + |x_n|)^{-2\beta} \int dx_n\ |\int d\hat{x}\ e^{-i\xi.\hat{x}} \cdot$$

$$\cdot \left[\frac{1 + |x_n|}{1 + |x|}\right]^\beta f(\hat{x},x_n)|^2]^{\frac{1}{2}} .$$

Integrating over $d\xi$ and using the unitarity of the Fourier transformation in $L^2(\mathbb{R}^{n-1})$, we obtain, with $c_\beta < \infty$:

$$\int d\xi\ \ |(F(1 + |Q|)^{-\beta} f)(\xi,\lambda)|^2$$

$$\leq c_\beta^2 \int dx_n \int d\hat{x} \left[\frac{1 + |x_n|}{1 + |x|}\right]^{2\beta} |f(\hat{x},x_n)|^2 \leq c_\beta^2 \|f\|^2 ,$$

since $1+|x_n| \leq 1 + |x|$. Hence $\|\hat{T}_\beta(\lambda)f\| \leq c_\beta \|f\|$, which proves the boundedness of $\hat{T}_\beta(\lambda)$.

The Hölder continuity of $T_\beta(\lambda)$ follows similarly. One has

$$|(\hat{T}_\beta(\lambda)f)(\xi) - (\hat{T}_\beta(\mu)f)(\xi)|$$

$$\leq (2\pi)^{-\frac{n}{2}} \int dx_n (1 + |x_n|)^{-\beta} |e^{-i\lambda x_n} - e^{-i\mu x_n}| \cdot$$

$$\cdot \left| \int d\hat{x}\, e^{-i\xi.\hat{x}} \left[\frac{1 + |x_n|}{1 + |x|}\right]^\beta f(\hat{x}, x_n)\right|.$$

Here one use the following inequality, valid for $a,b \in \mathbb{R}$ and $\nu \in [0,1]$:

$$|e^{-ia} - e^{-ib}| \leq 2^{1-\nu} |a-b|^\nu .$$

We take $0 < \nu < \min(\beta - \frac{1}{2}, 1)$ and obtain

$$\|\hat{T}_\beta(\lambda)f - \hat{T}_\beta(\mu)f\|^2 \leq c_\nu |\mu-\lambda|^{2\nu} \int dx_n \frac{|x_n|^{2\nu}}{(1 + |x_n|)^{2\beta}} \|f\|^2 .$$

Our choice of ν ensures the finiteness of the integral over dx_n, and the Hölder continuity of $T_\beta(\lambda)$ is proven. ∎

The restriction of a function to a sphere will be locally controlled by mapping a part of the sphere into a hyperplane. The necessary estimate is given by the next Lemma.

LEMMA 8.7 : Let Δ_1, Δ_2 be open subsets of $\mathbb{R}^n$, Δ_0 a compact subset of Δ_2 and η a C^∞-diffeomorphism of Δ_1 onto Δ_2. Set $\mathcal{D}_0 = \{f \mid \tilde{f} \in C_0^\infty(\Delta_0)\}$ and, for $f \in \mathcal{D}_0$, define f_η by $\tilde{f}_\eta = \tilde{f} \circ \eta$. Then, for each $\beta \geq 0$ there is a constant c_β such that

$$\|(1 + |Q|)^{\beta} f_{\eta}\| \leq c_{\beta}\|(1 + |Q|)^{\beta} f\| \quad \forall f \in \mathcal{D}_0 \ . \tag{8.18}$$

Proof :

(i) Assume first that $0 < \beta < 1$. By using the unitarity of Fourier transformation (with respect to the variable p), we find that

$$\begin{aligned}
&\iint dk\, dp\, |\tilde{g}(k) - \tilde{g}(p)|^2\, |k-p|^{-n-2\beta} \\
&= \iint dk\, dp\, |\tilde{g}(k+p) - \tilde{g}(p)|^2\, |k|^{-n-2\beta} \\
&= \iint dk\, dx\, |e^{-ik.x} - 1|^2\, |g(x)|^2\, |k|^{-n-2\beta} \ .
\end{aligned}$$

The integral over dk may be calculated explicitly. It is equal to $m_{\beta}|x|^{2\beta}$ for some constant m_{β} ($m_{\beta} < \infty$ since $0 < \beta < 1$). Hence we have the identity

$$\begin{aligned}
&m_{\beta}^{-1} \iint dk\, dp\, |\tilde{g}(k) - \tilde{g}(p)|^2\, |k-p|^{-n-2\beta} + \|g\|^2 \\
&= \int dx\, |g(x)|^2\, (1 + |x|^{2\beta}) = \|(1 + |Q|^{2\beta})^{\frac{1}{2}} g\|^2 .
\end{aligned} \tag{8.19}$$

Let Δ_0' be another compact subset of Δ_2 containing Δ_0 in its interior, and $\delta = \text{dist}\,(\Delta_0, \mathbb{R}^n\backslash\Delta_0') > 0$. If $f \in \mathcal{D}_0$, we have

$$\begin{aligned}
&\iint dk\, dp\, |\tilde{f}(k) - \tilde{f}(p)|^2\, |k-p|^{-n-2\beta} \\
&= \iint_{\Delta_0'\times\Delta_0'} dk\, dp\, |\tilde{f}(k) - \tilde{f}(p)|^2\, |k-p|^{-n-2\beta} \\
&+ \iint_{\mathbb{R}^n\backslash\Delta_0'\times\Delta_0} dk\, dp\, |\tilde{f}(k) - \tilde{f}(p)|^2\, |k-p|^{-n-2\beta} \qquad (8.20) \\
&+ \iint_{\Delta_0\times\mathbb{R}^n\backslash\Delta_0'} dk\, dp\, |\tilde{f}(k) - \tilde{f}(p)|^2\, |k-p|^{-n-2\beta} \ ,
\end{aligned}$$

since the integrand is zero in the remaining domain.

We shall also use the inequalities ($\beta,\gamma > 0$)

$$2^{-\beta}(1 + |x|)^{\beta} \leq (1 + |x|^{2\beta})^{\frac{1}{2}} \leq \sqrt{2}\,(1 + |x|)^{\beta} \tag{8.21}$$

and

$$2^{-\gamma}(1 + |x|)^{\gamma} \leq (1 + |x|^{2})^{\frac{\gamma}{2}} \leq 2^{\frac{\gamma}{2}}\,(1 + |x|)^{\gamma}\ . \tag{8.22}$$

(ii) Let $f \in \mathcal{D}_0$. From (8.21) and (8.19) we obtain

$$\|(1 + |Q|)^{\beta} f_{\eta}\|^{2} \leq 2^{2\beta}\|(1 + |Q|^{2\beta})^{\frac{1}{2}}\, f_{\eta}\|^{2}$$

$$= 2^{2\beta}\,\|f_{\eta}\|^{2} + 2^{2\beta}\, m_{\beta}^{-1} \iint dk\, dp\, |(\tilde{f}\circ\eta)(k) - (\tilde{f}\circ\eta)(p)|^{2}\, |k-p|^{-n-2\beta}.$$

In the last integral we use (8.20), i.e. we decompose the integral into three terms, over the domaine $(\eta^{-1}\,\Delta_0') \times (\eta^{-1}\,\Delta_0')$, $(\mathbb{R}^{n}\backslash\eta^{-1}\,\Delta_0') \times (\eta^{-1}\Delta_0)$ and $(\eta^{-1}\Delta_0) \times (\mathbb{R}^{n}\backslash\eta^{-1}\,\Delta_0')$. In the first domain we make the change of variables $k' = \eta(k)$, $p' = \eta(p)$ and use the fact that $|\eta(k) - \eta(p)| \leq c'\,|k-p|$ to obtain

$$\iint_{(\eta^{-1}\Delta_0')\times(\eta^{-1}\Delta_0')} dk\, dp\, |(\tilde{f}\circ\eta)(k) - (\tilde{f}\circ\eta)(p)|^{2}\, |k-p|^{-n-2\beta}$$

$$\leq (c')^{n+2\beta} \iint_{\Delta_0'\times\Delta_0'} dk'\, dp'\, |\tilde{f}(k') - \tilde{f}(p')|^{2}\, |k'-p'|^{-n-2\beta}\, |\tfrac{dk}{dk'}|\, |\tfrac{dp}{dp'}|$$

$$\leq (c')^{n+2\beta}\, (c'')^{2}\, m_{\beta}\, \|(1 + |Q|^{2\beta})^{\frac{1}{2}}\, f\|^{2}\ ,$$

where c'' is an upper bound for the Jacobian $|\frac{dk}{dk'}|$ in Δ_o'.

In the second domain, $(\tilde{f}\circ\eta)(k) = 0$, and the integral over dk is majorized uniformly in $p \in \Delta_0$, by a constant d_β depending on Δ_0 and Δ_0'. Hence the integral over the second domain is less than $c'' d_\beta \|f\|^2$. The same bound applies to the third integral. We thus have

$$\|(1 + |Q|)^\beta f_\eta\|^2 \le 2^{2\beta} m_1'' \|f\|^2 + 2^{2\beta}(c')^{n+2\beta} (c'')^2 \cdot$$

$$\cdot \|(1 + |Q|^{2\beta})^{\frac{1}{2}} f\|^2 + 2^{2\beta+1} c'' d_\beta m_\beta^{-1} \|f\|^2$$

$$\le M_\beta \|(1 + |Q|^{2\beta})^{\frac{1}{2}} f\|^2 ,$$

for a suitable $M_\beta < \infty$. By combining this with the second inequality in (8.21), we obtain (8.18) for $0 < \beta < 1$. For $\beta = 0$ one gets (8.18) by a simple change of variables.

Using the fact that $(\widetilde{Q_j f})(k) = i \dfrac{\partial \tilde{f}(k)}{\partial k_j}$, one obtains by a similar calculation as above that $(0 \le \beta < 1)$

$$\|(1 + |Q|)^\beta Q_i f_\eta\|^2 \le \sum_{j=1}^n c_\beta' \|(1 + |Q|)^\beta Q_j f\|^2 . \tag{8.23}$$

(iii) One has

$$\|(1 + |Q|^2)^{\frac{1}{2}} g\|^2 = (g, g + |Q|^2 g) = \|g\|^2 + \sum_{j=1}^n \|Q_j g\|^2 . \tag{8.24}$$

Thus

$$\|(1 + |Q|^2)^{\frac{(\beta+1)}{2}} f_\eta\|^2 = \|(1 + |Q|^2)^{\frac{\beta}{2}} f_\eta\|^2 + \sum_{j=1}^n \|(1 + |Q|^2)^{\frac{\beta}{2}} Q_j f_\eta\|^2 .$$

Together with (8.22) this implies that, for $\gamma = \beta+1$,

$$\|(1 + |Q|)^{\beta+1} f_\eta\|^2 \le 2^{2\gamma} \|(1 + |Q|^2)^{\frac{(\beta+1)}{2}} f_\eta\|^2$$

$$\le 2^{3\gamma}\|(1 + |Q|)^\beta f_\eta\|^2 + 2^{3\gamma} \sum_{j=1}^{n} \|(1 + |Q|)^\beta Q_j f_\eta\|^2 . \qquad (8.25)$$

Assume $0 \le \beta < 1$. Then we obtain from (8.25), (8.18), (8.23) and (8.24) that

$$\|(1 + |Q|)^{\beta+1} f_\eta\|^2 \le 2^{3\gamma} c_\beta^2 \|(1 + |Q|)^\beta f\|^2 + 2^{3\gamma} \sum_{j=1}^{n} n\, c'_\beta \cdot$$

$$\cdot \|(1 + |Q|)^\beta Q_j f\|^2 \le c_{\beta+1}\|(1 + |Q|)^{\beta+1} f\|^2$$

for some suitable $c_{\beta+1}$, i.e. (8.18) holds for $1 \le \beta < 2$. Repeating this argument, one finds that (8.18) holds for all $\beta \ge 0$. ∎

LEMMA 8.8 : Let $\phi \in S(\mathbb{R}^n)$ and $\nu \ge 0$. Then

$$(1 + |Q|)^\nu \phi(P)(1 + |Q|)^{-\nu} \in B(L^2(\mathbb{R}^n)) .$$

Proof : We shall use the inequality ($u,v \in \mathbb{R}^n$)

$$1 + |u+v| \le 1 + |u| + |v| \le (1 + |u|)(1 + |v|) . \qquad (8.26)$$

We denote by F_m the multiplication operator by the characteristic function of the ball $B_m = \{x \in \mathbb{R}^n \mid |x| \le m\}$.

The operator $(1 + |Q|)^\nu \phi(P)(1 + |Q|)^{-\nu}$ is an integral operator with kernel K(x,y) given by

$$K(x,y) = (2\pi)^{-\frac{n}{2}} (1 + |x|)^{\nu} \tilde{\phi}(y-x)(1 + |y|)^{-\nu} .$$

Now, using first (8.26) with u = x-y, v = y, and then the Schwarz inequality we get

$$\iint dx\, dy\, |\overline{f(x)}\, (1 + |x|)^{\nu} \tilde{\phi}(y-x)(1 + |y|)^{-\nu}\, g(y)|$$

$$\leq \iint dx\, dy\, |\overline{f(x)}\, g(y)|(1 + |y-x|)^{\nu}\, |\tilde{\phi}(y-x)|$$

$$\leq [\iint dx\, dy\, |f(x)|^2 (1 + |y-x|)^{\nu}\, |\tilde{\phi}(y-x)|]^{\frac{1}{2}} \cdot$$

$$\cdot [\iint dx\, dy\, |g(y)|^2 (1 + |y-x|)^{\nu}\, |\tilde{\phi}(y-x)|]^{\frac{1}{2}}$$

$$= \|f\|\, \|g\|\, \|h\|_1 ,$$

where $h(u) = (1 + |u|)^{\nu} \tilde{\phi}(u)$. Notice that the L^1-norm of h is finite, since $\tilde{\phi} \in S(\mathbb{R}^n)$.

For $g \in L^2(\mathbb{R}^n)$, define $g_m = (1 + |Q|)^{\nu} F_m\, \phi(P)(1 + |Q|)^{-\nu} g$, which is also in $L^2(\mathbb{R}^n)$ since $(1 + |Q|)^{\nu} F_m$ is bounded. Taking $f = g_m$ in the above inequality, we find that

$$\|g_m\|^2 = (2\pi)^{-\frac{n}{2}} \iint dx\, dy\, \overline{g_m(x)}(1 + |x|)^{\nu} \tilde{\phi}(y-x)(1 + |y|)^{-\nu} g(y)$$

$$\leq (2\pi)^{-\frac{n}{2}} \|g_m\|\, \|g\|\, \|h\|_1 . \tag{8.27}$$

If $\|g_m\| = 0$ for all m, we have $(1 + |Q|)^{\nu} \phi(P)(1 + |Q|)^{-\nu} g = 0$.

If $\|g_m\| \neq 0$ for some m, we obtain from (8.27) that $\|g_m\| \leq (2\pi)^{-\frac{n}{2}} \|h\|_1 \|g\|$, i.e.

$$\int_{|x|\leq m} dx\ (1 + |x|)^{2\nu}\ |(\phi(P)(1 + |Q|)^{-\nu} g)(x)|^2$$

$$\leq (2\pi)^{-n}\ \|h\|_1^2\ \|g\|^2 \quad .$$

Letting $m \to \infty$, we find that

$$\|(1 + |Q|)^{\nu}\ \phi(P)(1 + |Q|)^{-\nu}\ g\|^2 \leq (2\pi)^{-n}\ \|h\|_1^2\ \|g\|^2 \ . \ \blacksquare$$

PROPOSITION 8.9 : Let $n \geq 2$, $\beta > \frac{1}{2}$ and $\phi(x) = (1 + |x|)^{-\beta}$. Then

(a) For each $\lambda > 0$ the operator $\hat{M}_\phi(\lambda)$, defined by (8.1) is bounded, and the function $\lambda \mapsto M_\phi(\lambda)$ is locally Hölder continuous on $(0,\infty)$ in the operator norm, for each Hölder exponent $\gamma < \min\ (\beta - \frac{1}{2}, 1)$.

(b) $M_\phi(\lambda)^*$ satisfies all statements (a) - (d) of Proposition 8.4 if Δ is a closed interval in $(0,\infty)$ and $\psi(\lambda) = 0$ in a neighbourhood of $\lambda = 0$.

Proof : Let $f \in \mathcal{S}(\mathbb{R}^n)$. Let $\lambda > 0$ be fixed, let $0 < \delta < \lambda$ and let $S_{\lambda,\delta}$ be the following spherical shell :

$$S_{\lambda,\delta} = \{k \in \mathbb{R}^n \mid \lambda-\delta < |k|^2 < \lambda+\delta\} \ .$$

Let $B_1,\ldots,B_r$ be open balls such that $S_{\lambda,\delta} \subset \bigcup_i B_i$ and such that $0 \notin \bar{B}_i$ for each i. Let $\psi_1,\ldots,\psi_r$ be such that $\psi_i \in C_0^\infty(B_i)$ and $\sum_{i=1}^r \psi_i(k) = 1$ for $k \in S_{\lambda,\delta}$. Then for $\mu \in (\lambda-\delta,\lambda+\delta)$, $\hat{M}_\phi(\mu) = \sum_{i=1}^r \hat{M}_\phi^i(\mu)$, with

$$(\hat{M}^i_\phi(\mu)f)(\omega) = \psi_i(\sqrt{\mu}\omega)\ (\mathcal{U}_o\phi f)_\mu\ (\omega)$$

$$= (\mathcal{U}_o\ \psi_i(P)(1 + |Q|)^{-\beta}\ f)_\mu\ (\omega)\ .$$

It suffices to show that each $\hat{M}^i_\phi$ has the properties stated in the Proposition. Let $S_\mu = \{k \mid k^2 = \mu\}$. Let $\eta_{i,o}$ be a C^∞-diffeomorphism from a sufficiently large subset Σ_i of S_1 onto some open subset of $\mathbb{R}^{n-1}$, and define $\eta_i : B_i \to \mathbb{R}^n$ by

$$\eta_i : \sqrt{\mu}\omega \mapsto (\eta_{i,o}(\omega),\mu)\ , \qquad \text{for } \sqrt{\mu}\omega \in B_i\ .$$

Let c_i be an upper bound for the Jacobian of $\eta^{-1}_{i,o}$ in the domain $\eta_i\ \Sigma_i$. Define $h_i \in \mathcal{S}(\mathbb{R}^n)$ by

$$\tilde{h}_i(k) = (\mathcal{U}_o\ \psi_i(P)\ \phi(Q)f)_{k_n^2}\ (\eta^{-1}_{i,o}(k_1,\dots,k_{n-1}))$$

if $k_n^2 \cdot \eta^{-1}_{i,o}(k_1,\dots,k_{n-1}) \in B_i$, and $\tilde{h}_i(k) = 0$ otherwise. By a change of variables $\omega \mapsto (\xi_1,\dots,\xi_{n-1}) = \eta_{i,o}(\omega)$ and using Lemma 8.6, we now obtain

$$\begin{aligned} \|\hat{M}^i_\phi(\mu)f\|^2 &= \int d\omega\ |(\mathcal{U}_o\ \psi_i(P)\ \phi(Q)f)_\mu\ (\omega)|^2 \\ &\le c_i \int d\xi_1 \dots d\xi_{n-1}\ |\tilde{h}_i(\xi_1,\dots,\xi_{n-1},\mu)|^2 \\ &= c_i\ \|\hat{T}_\beta(\mu)(1 + |Q|)^\beta\ h_i\|^2 \\ &\le c_i\ \|\hat{T}_\beta(\mu)\|^2\ \|(1 + |Q|)^\beta\ h_i\|^2\ . \end{aligned}$$

Since by Lemmas 8.7 and 8.8

$$\|(1+|Q|)^{\beta}h_i\| \le c_\beta \|(1+|Q|)^{\beta}\ \psi_i(P)(1+|Q|)^{-\beta}f\ \|$$

$$\le c_\beta\ d_i\|f\| \qquad (d_i < \infty)\ ,$$

we conclude that $\hat{M}^i_\phi(\mu)$ is bounded. A similar calculation leads to the Hölder continuity of $M^i_\phi(\mu)$ in μ. The proof of (b) is identical with that of Proposition 8.4. ∎

Remark : If instead of $\phi(x) = c(1+|x|)^{-\beta}$, we assume $|\phi(x)| \le c(1+|x|)^{-\beta}$, i.e. $\phi(x) = (1+|x|)^{-\beta}\ \psi(x)$ with $\psi \in L^\infty(\mathbb{R}^n)$, then all results of Proposition 8.9 are also true. This follows simply by observing that

$$M_{AB}(\lambda)\ f = M_A(\lambda)\ B\ f \tag{8.28}$$

(take $A = (1+|Q|)^{-\beta}$ and $B = \psi(Q)$).

From Proposition 8.5 and 8.9, we immediately obtain :

PROPOSITION 8.10 : Let $\phi_1,\phi_2 : \mathbb{R}^n \to \mathbb{C}$ be such that $|\phi_i(x)| \le c_i(1+|x|)^{-\beta_i}$ with $\beta_i > \frac{1}{2}$, $n \ge 2$. Then for each $\lambda > 0$ the following limits exist :

$$\underset{\varepsilon \to +0}{\text{u-lim}}\ \phi_1(Q)^*\ (K_o - \lambda \mp i\varepsilon)^{-1}\ \phi_2(Q) \equiv Z_{\lambda \pm io}\ .$$

The convergence is uniform on each compact subset of $(0,\infty)$ and

$$Z_{\lambda+io} - Z_{\lambda-io} = 2\pi i\ M_{\phi_1}(\lambda)^*\ M_{\phi_2}(\lambda)\ .$$

We also get new results on wave operators :

PROPOSITION 8.11 : Let $n \geq 3$, and let $V : \mathbb{R}^n \to \mathbb{R}$ be such that, for some $\delta > 0$, $(1 + |x|)^{1+\delta} V(x) = W_1(x) + W_2(x)$ with $W_1(x) \in L^{p_1}(\mathbb{R}^n) \cap L^{p_2}(\mathbb{R}^n)$, $p_1 < n < p_2$ and $|W_2(x)| \leq c(1 + |x|)^{-\frac{1}{2}-\nu}$ $(\nu > 0)$. Then $\Omega_\pm$ exist and are complete.

The proof is very similar to that of Proposition 7.8, using Lemma 7.7 and Proposition 8.10. We refrain from writing out the details.

From Proposition 6.13 we know that the wave operators exist under weaker decay assumptions on the potential $V(x)$ than those of Proposition 8.11, namely if $|V(x)| \leq c(1 + |x|)^{-1-\nu}$, $\nu > 0$. Asymptotic completeness for such potentials can be also proved, but this proof requires a refinement of the method given in these notes, for instance the theory of smooth operators introduced by (Kato, [26]). We do not want to develop here the theory of smooth operators. However, we point out that H_0-smoothness and local H-smoothness of $|V|^{1/2}$ follow from Proposition 8.10, Lemma 7.5 and the second resolvent equation.

These results show that the behaviour of $V(x)$ at infinity, which was crucial for the existence of the wave operators, is of little importance for their completeness. Completeness depends on the local behaviour of $V(x)$. We illustrate this by the following examples.

Suppose $V(x) = V(r)$, $r = |x|$, i.e. that V is a spherically symmetric function.

(a) If $V(r)$ is very rapidly oscillating near $r = 0$, then $\Omega_\pm$ need not be complete. An explicit example has been given by Pearson (Comm. Math. Phys. 40, 125, (1975)).

(b) Decomposition of configuration space :

write

$$\mathcal{H} = L^2(\mathbb{R}^3) = \mathcal{H}_1 \oplus \mathcal{H}_2 \equiv L^2(S_1) \oplus L^2(\mathbb{R}^3 \backslash S_1)$$

where

$$S_1 = \{x \mid |x| < 1\} \quad . \quad \text{We set}$$

$$\hat{H}_1 = (-\Delta+V) \qquad D(\hat{H}_1) = C_0^\infty(S_1)$$

$$\hat{H}_2 = (-\Delta+V) \qquad D(\hat{H}_2) = C_0^\infty(\mathbb{R}^3\setminus S_1) .$$

(the functions in $D(\hat{H}_1)$ vanish in the neighbourhood of the unit sphere and have support in the unit ball).

Assume that $\hat{H}_1$ and $\hat{H}_2$ are essentially self-ajoint in $\mathcal{H}_1$ and $\mathcal{H}_2$ respectively. Let H_1 and H_2 be their closures. Then $H = H_1 \oplus H_2$ is self-adjoint. If $V(x) \to 0$ as $|x| \to \infty$ faster than $|x|^{-1}$, $\Omega_\pm$ exist.

Let F be the projection with range $\mathcal{H}_1$. We have

$$e^{iHt}\, e^{-itK_0} = e^{itH_1}\, F\, e^{-itK_0} + e^{itH_2}\, (I - F)\, e^{-itK_0} .$$

By Corollary 6.11(b) it follows that $\operatorname*{s-lim}_{t\to\pm\infty} e^{itH_1}\, F\, e^{-itK_0} = 0$. Hence

$$\Omega_\pm = \operatorname*{s-lim}_{t\to\pm\infty} e^{itH_2}\, (I - F)\, e^{-itK_o}$$

$$= \operatorname*{s-lim}_{t\to\pm\infty} (I - F)\, e^{itH_2}\, e^{-itK_0} .$$

Therefore (range $\Omega_\pm$) $\subseteq \mathcal{H}_2$.

On the other hand $\mathcal{H}_{ac}(H) = \mathcal{H}_{ac}(H_1) \oplus \mathcal{H}_{ac}(H_2)$, so that (range $\Omega_\pm$) $\neq \mathcal{H}_{ac}(H)$ as soon as $\mathcal{H}_{ac}(H_1) \neq 0$. It is known that there are functions V(r) such that $\hat{H}_1$ and $\hat{H}_2$ are essentially self-adjoint and $\mathcal{H}_{ac}(H_1) \neq 0$ [e.g. $V(r) \geq (r-1)^{-2}$

for $1 < r \leq 2$, $V(r)$ rapidly oscillating with increasing amplitude as $r \to 1$ from below, see [31]],

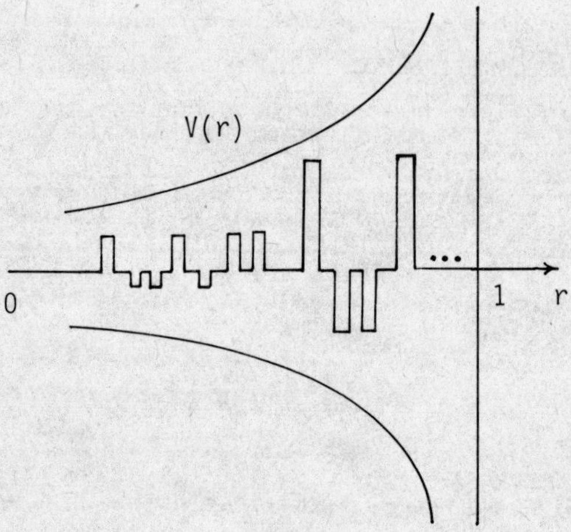

as $r \to 1-$:
the amplitude diverges sufficiently rapidly. The length of oscillation decreases to zero sufficiently rapidly.

9 The singular spectrum

In this chapter we discuss two points :

(i) we prove that the continuous spectrum of a class of Schrödinger operators is absolutely continuous,

(ii) we give some bounds on the number of eigenvalues of Schrödinger operators. We begin by the first point.

PROPOSITION 9.1 : Let $V = AB$, where A is multiplication in $L^2(\mathbb{R}^n)$ by $(1 + |x|)^{-\alpha}$, $\alpha > \frac{1}{2}$, B is multiplication by a function $B(x)$ and B is K_0-bounded. Assume that $W_z = B(K_0-z)^{-1} A$ satisfies the hypotheses of Lemma 7.5. Let $\lambda \in \Gamma_0^+ \cap (0,\infty)$ [respectively $\lambda \in \Gamma_0^- \cap (0,\infty)$], and let $f_+ \neq 0$ [respectively $f_- \neq 0$] be such that

$$(I + W_{\lambda+io})f_+ = 0 \qquad [\text{resp. } (I + W_{\lambda-io})f_- = 0].$$

Then

$$M_A(\lambda)f_+ = 0 \qquad [\text{resp. } M_A(\lambda)f_- = 0] .$$

Proof : Define

$$Z_{\lambda\pm io} = \underset{\varepsilon\to\pm 0}{\text{u-lim}} \; A(K_0-\lambda\mp i\varepsilon)^{-1} A .$$

Notice that the limits exist by Proposition 8.5 and Proposition 8.9. Let C be the self-adjoint multiplication operator by the function $C(x) = B(x)\, A(x)^{-1}$. We have for each $g \in L^2(\mathbb{R}^n)$:

$$Z_{\lambda\pm i\varepsilon}\, g \in D(C) \quad , \quad C\, Z_{\lambda\pm i\varepsilon}\, g = W_{\lambda\pm i\varepsilon}\, g \,,$$

$$\underset{\varepsilon \to +0}{\text{s-lim}}\; Z_{\lambda\pm i\varepsilon}\, g = Z_{\lambda\pm io}\, g$$

and

$$\underset{\varepsilon \to +0}{\text{s-lim}}\; C\, Z_{\lambda\pm i\varepsilon}\, g = W_{\lambda\pm io}\, g.$$

Since C is closed, this implies that $Z_{\lambda\pm io}\, g \in D(C)$ and $C\, Z_{\lambda\pm io}\, g = W_{\lambda\pm io}\, g$.

By Proposition 8.5 (b), and since $Z^{*}_{\lambda - io} = Z_{\lambda+io}$:

$$(M_A(\lambda)f_+ \,, M_A(\lambda)f_+) = (f_+,\, M_A(\lambda)^{*}\, M_A(\lambda)f_+)$$

$$= \frac{1}{2\pi i}\,(f_+,[Z_{\lambda+io} - Z_{\lambda-io}]f_+)$$

$$= \frac{1}{2\pi i}\,(f_+,\, Z_{\lambda+io}f_+) - \frac{1}{2\pi i}\,\overline{(f_+,\, Z^{*}_{\lambda-io}f_+)}$$

$$= \frac{1}{\pi}\,\mathrm{Im}(f_+,\, Z_{\lambda+io}f_+) = -\,\frac{1}{\pi}\,\mathrm{Im}(W_{\lambda+io}f_+ \,, Z_{\lambda+io}f_+)$$

$$= -\,\frac{1}{\pi}\,\mathrm{Im}(C\, Z_{\lambda+io}f_+ \,,\, Z_{\lambda+io}f_+) \,.$$

Since C is self-adjoint, $(g,Cg) = (Cg,g) = \overline{(g,Cg)}$ for each $g \in D(C)$, i.e. $\mathrm{Im}(g,Cg) = 0$. Hence $\|M_A(\lambda)f_+\|^2 = 0$, i.e. $M_A(\lambda)f_+ = 0$. ∎

<u>COROLLARY 9.2</u> : Assume the hypotheses of Proposition 9.1. Then $\Gamma_o^+ \cap (0,\infty) = \Gamma_o^- \cap (0,\infty)$.

<u>Proof</u> : Let $\lambda > 0$, $\lambda \in \Gamma_o^+$. Let f_+ be such that $(I + W_{\lambda+io})f_+ = 0$. Then by Proposition 9.1

$$(I + W_{\lambda-io})f_+ = (I + W_{\lambda+io})f_+ - (W_{\lambda+io} - W_{\lambda-io})f_+$$

$$= 0 - C(Z_{\lambda+io} - Z_{\lambda-io})f_+$$

$$= - 2\pi i\ C(M_A(\lambda)^* M_A(\lambda))f_+ = 0 .$$

Hence $I + W_{\lambda-io}$ is not invertible, i.e. $\lambda \in \Gamma_o^-$, which implies that $(0,\infty) \cap \Gamma_o^+ \subseteq (0,\infty) \cap \Gamma_o^-$. Similarly one proves the opposite inclusion, giving $(0,\infty) \cap \Gamma_o^+ = (0,\infty) \cap \Gamma_o^-$. ■

PROPOSITION 9.3 : Assume the hypotheses of Proposition 9.1, and, in addition, that the Hölder index γ for $M_A(\mu)$ satifies $\gamma > \frac{1}{2}$ (i.e. $\alpha > 1$ in Proposition 9.1). Define $g_\pm$ by

$$(\mathcal{U}_o\ g_\pm)_\mu = (\mu-\lambda)^{-1}(\mathcal{U}_o\ Af_\pm)_\mu$$

$$= (\mu-\lambda)^{-1}\ M_A(\mu)f_\pm \in L^2(S^{n-1}) .$$

Then

(a) $$g_\pm \in D(H) \subset L^2(\mathbb{R}^n)$$

and

$$Hg_\pm = \lambda g_\pm \quad .$$

(b) $$\mathcal{H}_{sc}(H) = \{0\}.$$

Proof : First notice that (see also Problem 14.1)

$$\int_0^\infty d\mu \ \|M_A(\mu)f_\pm\|^2 = \int_0^\infty d\mu \ \|(\mathcal{U}_0 Af_\pm)_\mu\|^2 = \|Af_\pm\|^2 < \infty.$$

(i) $$\|g_\pm\|^2 = \int_0^\infty d\mu \ \|(\mathcal{U}_0 g_\pm)_\mu\|^2$$

$$= \int_{|\mu-\lambda|>\frac{\lambda}{2}} d\mu \ \|(\mathcal{U}_0 g_\pm)_\mu\|^2 + \int_{|\mu-\lambda|<\frac{\lambda}{2}} d\mu \ \|(\mathcal{U}_0 g_\pm)_\mu\|^2 \ .$$

Now

$$\int_{|\mu-\lambda|>\frac{\lambda}{2}} d\mu \ \|(\mathcal{U}_0 g_\pm)_\mu\|^2 \le (\tfrac{\lambda}{2})^{-2} \int_0^\infty d\mu \ \|M_A(\mu)f_\pm\|^2$$

$$= 4\lambda^{-2} \ \|Af_\pm\|^2 < \infty ,$$

since $\lambda > 0$. For $\mu \in (\frac{1}{2}\lambda, \frac{3}{2}\lambda)$ we get from Proposition 9.1

$$\|M_A(\mu)f_\pm\| = \|[M_A(\mu) - M_A(\lambda)]f_\pm\| \le c_\lambda \ |\mu-\lambda|^\gamma \ \|f_\pm\| \ .$$

Hence

$$\int_{|\mu-\lambda|<\frac{\lambda}{2}} d\mu \ \|(\mathcal{U}_0 g_\pm)_\mu\|^2 \le \int_0^{2\lambda} d\mu \ |\mu-\lambda|^{-2} \ c_\lambda^2 \ |\mu-\lambda|^{2\gamma} \ \|f_\pm\|^2 ,$$

which is finite since $2\gamma > 1$. Hence $g_\pm \in L^2(\mathbb{R}^n)$.

(ii) We have

$$\mu(\mathcal{U}_0 g_\pm)_\mu = (\mathcal{U}_0 Af_\pm)_\mu + \lambda(\mathcal{U}_0 g_\pm)_\mu \ .$$

Since each term defines a vector in $L^2(\mathbb{R}^n)$ and

$$(\mathcal{U}_0 K_0 g_\pm)_\mu = \mu(\mathcal{U}_0 g_\pm)_\mu ,$$

we have

$$g_\pm \in D(K_0) = D(H)$$

and

$$K_0 g_\pm = A f_\pm + \lambda g_\pm . \tag{9.1}$$

(iii) Let $g_\pm^\varepsilon = (K_0 - \lambda \mp i\varepsilon)^{-1} A f_\pm$, $\varepsilon > 0$. One has $g_\pm^\varepsilon \in D(K_0) \subseteq D(B)$ and

$$\operatorname*{s-lim}_{\varepsilon \to 0} B g_\pm^\varepsilon = W_{\lambda \pm i0} f_\pm .$$

Furthermore

$$\| g_\pm^\varepsilon - g_\pm \|^2 = \int |(\mu - \lambda \mp i\varepsilon)^{-1} - (\mu - \lambda)^{-1}|^2 \, \|(\mathcal{U}_0 A f_\pm)_\mu\|^2 \, d\mu ,$$

which converges to zero as $\varepsilon \to 0$ by the Lebesgue dominated convergence theorem, since the integrand converges to zero almost everywhere, and is majorized uniformly in ε, by

$$|\mu - \lambda|^{-2} \, \|(\mathcal{U}_0 A f_\pm)_\mu\|^2 ,$$

which is in $L^1((0,\infty), d\mu)$ by (i) .
Hence s-lim $g_\pm^\varepsilon = g_\pm$ and s-lim $B g_\pm^\varepsilon = W_{\lambda \pm i0} f$ as $\varepsilon \to 0$. Since B is closed, we have $g_\pm \in D(B)$ [which we already know since $g_\pm \in D(K_0) \subseteq D(B)$] and

$$B g_\pm = W_{\lambda \pm i0} f_\pm = - f_\pm . \tag{9.2}$$

(iv) From (9.1) and (9.2) we deduce that

$$Hg_\pm = K_0 g_\pm + Vg_\pm = Af_\pm + \lambda g_\pm + ABg_\pm$$

$$= Af_\pm + \lambda g_\pm - Af_\pm = \lambda g_\pm ,$$

which proves (a).

(v) Since $\sigma_{sc}(H) \subseteq \sigma_e(H) = [0,\infty)$ [see the end of chapter 4] and $\sigma_{sc}(H) \subseteq \Gamma_0$ by the proof of Proposition 7.6, each point $\lambda \neq 0$ in $\sigma_{sc}(H)$ belongs to $(0,\infty) \cap \Gamma_0$. Since eigenvectors of H belonging to different eigenvalues are orthogonal and $L^2(\mathbb{R}^n)$ is separable, H can have at most a countable number of eigenvalues. Since $\lambda \in (0,\infty) \cap \Gamma_0$ implies $\lambda \in \sigma_p(H)$, $(0,\infty) \cap \Gamma_0$ is at most a countable set, i.e. $(0,\infty) \cap \Gamma_0 = \bigcup_{i=1}^{\infty} \{\lambda_i\}$. Then, for $g \in L^2(\mathbb{R}^n)$:

$$E_{(0,\infty)\cap\Gamma_0}\, g = \sum_{i=1}^{\infty} E_{\{\lambda_i\}}\, g .$$

$E_{\{\lambda_i\}}\, g$ is either 0 or an eigenvector of H with eigenvalue λ_i. Thus $E_{(0,\infty)\cap\Gamma_0}\, g \in \mathcal{H}_p(H)$ for each g. Hence

$$\mathcal{H}_{sc}(H) \subseteq E_{(0,\infty)\cap\Gamma_0}\mathcal{H} \subseteq \mathcal{H}_p(H) .$$

Since $\mathcal{H}_{sc}(H) \perp \mathcal{H}_p(H)$, we have $\mathcal{H}_{sc}(H) = \{0\}$. ∎

Example 9.4 : Proposition 9.3 implies that $\mathcal{H}_{sc}(H) = \{0\}$ for example for the following type of potentials V :

$$V(x) \equiv A(x)\, B(x) = (1 + |x|)^{-1-\delta}\, [B_1(x) + B_2(x)],$$

with $|B_1(x)| \le c(1 + |x|)^{-\frac{1}{2}-\varepsilon}$ and $B_2 \in L^{p_1}(\mathbb{R}^n) \cap L^{p_2}(\mathbb{R}^n)$,

$2 \le p_1 < n < p_2$, $\delta > 0$, $\varepsilon > 0$. ($n \ge 3$).

We next give some further properties of the null spaces of $I + W_{\lambda \pm io}$. By Corollary 9.2, the null spaces of $I + W_{\lambda+io}$ and of $I + W_{\lambda-io}$ are identical if $\lambda > 0$, and will therefore be denoted by $N(\lambda)$. For $\lambda \le 0$, one has $W_{\lambda+io} = W_{\lambda-io} = W_\lambda$, so that we may again use the notation $N(\lambda)$ for the null space of $I + W_\lambda$.

If $\lambda \in \mathbb{R}$ is an eigenvalue, we denote by $M_{\{\lambda\}}$ the corresponding eigenspace, i.e. $M_{\{\lambda\}} = \{f \in D(H) \mid Hf = \lambda f\}$. One then has the following result :

<u>PROPOSITION 9.5</u> : Let V be as Proposition 9.3.

(a) If $\lambda \in \mathbb{R}$ is an eigenvalue of $H = K_0 + V$, then B maps $M_{\{\lambda\}}$ injectively into $N(\lambda)$. In particular $N(\lambda) \ne \{0\}$. Furthermore

$$\dim M_{\{\lambda\}} \le \dim N(\lambda) < \infty. \tag{9.3}$$

(b) If in addition $\lambda \ne 0$, B maps $M_{\{\lambda\}}$ onto $N(\lambda)$. Hence $\dim M_{\{\lambda\}} = \dim N(\lambda)$. Also, if $\lambda \ne 0$ and $N(\lambda) \ne \{0\}$, then λ is an eigenvalue of H.

<u>Proof</u> :

(i) Suppose $Hg = \lambda g$, $g \in D(H) \subseteq D(K_0) \subseteq D(B)$. Then

$$\begin{aligned}(I + W_{\lambda+io})\,Bg &= Bg + \operatorname*{s-lim}_{\varepsilon \to 0} B(K_0-\lambda-i\varepsilon)^{-1}\,Vg \\ &= Bg + \operatorname*{s-lim}_{\varepsilon \to 0} B(K_0-\lambda-i\varepsilon)^{-1}\,(Hg - K_0 g) \\ &= Bg - Bg - \operatorname*{s-lim}_{\varepsilon \to 0} (i\varepsilon)\,B(K_0-\lambda-i\varepsilon)^{-1}\,g\ .\end{aligned}$$

Hence, for each $f \in D(B)$:

$$(f, (I + W_{\lambda+io})Bg) = -i \lim_{\varepsilon\to 0} (Bf, \varepsilon(K_o-\lambda-i\varepsilon)^{-1} g) = 0 \quad (9.4)$$

since

$$\underset{\varepsilon\to 0}{\text{s-lim}}\ \varepsilon(K_o-\lambda-i\varepsilon)^{-1} = 0. \quad (9.5)$$

(9.4) shows that $(I + W_{\lambda+io})$ Bg is orthogonal to the dense set D(B), hence equal to zero, which shows that $Bg \in N(\lambda)$. Thus B maps $M_{\{\lambda\}}$ into $N(\lambda)$. The injectivity of this mapping is easily obtained : If $Hg = \lambda g$ and $Bg = 0$, then $K_o g = K_o g + ABg = Hg = \lambda g$, hence g is an eigenvector of K_o. Since K_o is spectrally absolutely continuous, we must have $g = 0$.

(ii) If $\lambda > 0$ and $N(\lambda) \neq \{0\}$, λ is an eigenvalue of H and B maps $M_{\{\lambda\}}$ onto $N(\lambda)$ by Proposition 9.3, in particular equation (9.2). If $\lambda < 0$, the analysis of Proposition 9.3 becomes much simpler : If $(I + W_\lambda)f = 0$, set $g = (K_o-\lambda)^{-1} Af$. Then $Bg = W_\lambda f = -f$, and consequently $(H-\lambda)g = (K_o-\lambda+AB)g = Af - Af = 0$. Again B is surjective from $M_{\{\lambda\}}$ to $N(\lambda)$.

Since B maps $M_{\{\lambda\}}$ injectively and surjectively onto $N(\lambda)$, one has

$$\dim M_{\{\lambda\}} = \dim N(\lambda) \quad \text{if } \lambda \neq 0 .$$

Finally we prove that $\dim N(\lambda) < \infty$. Assume the contrary that $\dim N(\lambda) = \infty$. Since $W_{\lambda+io}$ is compact, there is a sequence $\{T_n\}$ of finite rank operators such that $\|W_{\lambda+io} - T_n\| \to 0$ as $n \to \infty$.

Let $P(\lambda)$ be the orthogonal projection with range $N(\lambda)$. Then $P(\lambda)\, T_n P(\lambda)$ is a finite rank operator in $N(\lambda)$ for each n, and

$$\begin{aligned} &\|P(\lambda)\, W_{\lambda+io}\, P(\lambda) - P(\lambda)\, T_n\, P(\lambda)\| \\ &\quad \le \|P(\lambda)\|\, \|W_{\lambda+io} - T_n\|\, \|P(\lambda)\| \to 0 \quad \text{as } n \to \infty. \end{aligned} \quad (9.6)$$

Now, since $W_{\lambda+io} P(\lambda) = -P(\lambda)$, we have

$$P(\lambda) W_{\lambda+io} P(\lambda) = - P(\lambda) .$$

If $\dim N(\lambda) = \infty$, we may find for each n a vector f_n in $N(\lambda)$ such that $P(\lambda) T_n P(\lambda) f_n = 0$ and $\|f_n\| = 1$. Then

$$\|P(\lambda) W_{\lambda+io} P(\lambda) f_n - P(\lambda) T_n P(\lambda) f_n\| = \|P(\lambda) f_n\| = \|f_n\| = 1 \quad ,$$

which contradicts (9.6). ■

Proposition (9.5) shows that every $\lambda \in \Gamma_0 \setminus \{0\}$ is an eigenvalue of H of finite multiplicity, if V satisfies the hypotheses of Proposition 9.3. We shall now show, for a subset of this class of potentials, that $\dim E_{(-\infty,0)}\mathcal{H} < \infty$, in other words that the number of negative eigenvalues of H, multiplicities counted, is finite. This need not be the case for different V. For example, if $V(x) \leq 0$ for $|x| > R$, $V(.) \in L^2_{Loc}(\mathbb{R}^3)$, $\lim_{|x|\to\infty} V(x) = 0$ and $V(x) < - x^{-2}$ for $|x| > R$, then $H = K_0 + V$, acting in $L^2(\mathbb{R}^3)$ has an infinite number of negative eigenvalues. Since $\sigma_e(H) = [0,\infty)$, these eigenvalues accumulate at the point $\lambda = 0$.

As regards the positive eigenvalues, they are usually absent (although this fact has not been proved for the whole class of potentials considered here).
The proof that $\sigma_p(H) \cap (0,\infty) = \phi$ requires different methods from those used here (Jansen and Kalf, [24]), (A.M. Berthier, [11]). Results in this direction will be given in chapter X. What can be shown with the methods developed here is that the positive eigenvalues can accumulate at most at the point $\lambda = 0$ (Agmon, [1]) .

Let $H = K_0 + V$ with V K_0-compact. Then [see Chapter 4] H is bounded below and $\sigma_e(H) = \sigma_e(K_0) = [0,\infty)$. Let $\lambda_1 \leq \lambda_2 \leq \ldots \leq \lambda_N$ be an enumeration of the negative eigenvalues of H, multiplicity counted, and let e_k be the corresponding eigenvectors, i.e.

$$He_k = \lambda_k e_k$$

$$\|e_k\| = 1$$

$$(e_j, e_k) = 0 \quad \text{if} \quad j \neq k \quad .$$

N may be infinite, in which case the sequence $\{\lambda_k\}$ accumulates at $\lambda = 0$. The following theorem is a special case of the so-called *minimax principle* which gives an expression for λ_n.

<u>PROPOSITION 9.6</u> : Let H be as above. For $f_1, \ldots, f_{n-1} \in \mathcal{H}$, define

$$\rho(f_1, \ldots, f_{n-1}) = \inf_{\substack{g \in D(H),\ \|g\|=1 \\ g \perp f_k \text{ for } k=1,\ldots,n-1}} (g, Hg) \quad . \tag{9.7}$$

Let $\mu_1 = \inf\limits_{\substack{g \in D(H) \\ \|g\|=1}} (g, Hg)$, and for $n = 2, 3, \ldots$

$$\mu_n = \max_{f_1, \ldots, f_{n-1}} \rho(f_1, \ldots, f_{n-1}) \; .$$

Then

(a) $\mu_{n+1} \geq \mu_n$.

(b) if $n \leq N$, $\mu_n = \lambda_n$.

(c) If $n > N$, or if H has no negative eigenvalues, $\mu_n = 0$.

Proof :

(a) is evident from the definitions.

(b) Let $n \le N$. Consider the system of $(n-1)$ linear equations

$$\sum_{k=1}^{n} (f_j,e_k)\, x_k = 0 \ , \ j = 1,\ldots,n-1$$

for fixed $f_1,\ldots,f_{n-1}$. Let $\{x_1,\ldots,x_n\}$ be a non-trivial solution such that $\sum_{k=1}^{n} |x_k|^2 = 1$, and set

$$g = \sum_{k=1}^{n} x_k\, e_k \ .$$

Clearly $g \in D(H)$, $\|g\|^2 = \sum_{k=1}^{n} |x_k|^2 = 1$, $(f_j,g) = 0$ for $j = 1,\ldots,n-1$, and

$$(g,Hg) \ge \rho(f_1,\ldots,f_{n-1}) \ .$$

Also

$$(g,Hg) = \sum_{j,k} \bar{x}_j\, x_k\, (e_j,He_k) = \sum_{j,k} \bar{x}_j\, x_k\, \lambda_k\, \delta_{jk}$$

$$= \sum_{k=1}^{n} \lambda_k\, |x_k|^2 \le \lambda_n \sum_{k=1}^{n} |x_k|^2 = \lambda_n \ .$$

Thus

$$\lambda_n \ge \rho(f_1,\ldots,f_{n-1}) \ ,$$

hence

$$\mu_n = \max_{f_1,\ldots,f_{n-1}} \rho(f_1,\ldots,f_{n-1}) \le \lambda_n \ .$$

Now $\rho(e_1,\dots,e_{n-1}) \geq \lambda_n$ since $\{e_1,\dots,e_{n-1}\}^{\perp} \subseteq E_{[\lambda_n,\infty)}\mathcal{H}$,

so that $(g,Hg) \geq \lambda_n$ for each $g \perp \{e_1,\dots,e_{n-1}\}$ by the spectral theorem. Thus $\mu_n = \lambda_n$.

(c) If $n > N > 0$: $\rho(e_1,\dots,e_N,f_{N+1},\dots,f_{n-1}) \geq 0$ as above. If $a > 0$, then $\dim E_{[0,a]}\mathcal{H} = \infty$. Hence there exists $g \in E_{[0,a]}\mathcal{H}$ with $\|g\| = 1$, $g \perp \{e_1,\dots,e_n,f_{N+1},\dots,f_{n-1}\}$. Then $0 \leq (g,Hg) \leq a$, implying that

$$\rho(e_1,\dots,e_N,f_{N+1},\dots,f_{n-1}) = 0 .$$

A similar argument shows that for arbitrary $f_1,\dots,f_{n-1}$, $\rho(f_1,\dots,f_{n-1}) \leq 0$. Therefore

$$\mu_n = \rho(e_1,\dots,e_N,f_{N+1},\dots,f_{n-1}) = 0 .$$

If H has no negative eigenvalue, then the same reasoning gives $\rho(f_1,\dots,f_{n-1}) = 0$ for all $f_1,\dots,f_{n-1} \in \mathcal{H}$, hence $\mu_n = 0$. ∎

COROLLARY 9.7 : Let $H_i = K_0 + V_i$, $i = 1,2$, and assume that $V_1(x) \leq V_2(x)$ a.e. Then

(a) $$\dim E^{(1)}_{(-\infty,0)}\mathcal{H} \geq \dim E^{(2)}_{(-\infty,0)}\mathcal{H} ,$$

where $\{E^{(i)}_\lambda\}$ is the spectral family of H_i. In other words the number of negative eigenvalues of H_2, multiplicities counted, cannot exceed that of H_1.

(b) $$\mu^{(1)}_n \leq \mu^{(2)}_n \quad \text{for each } n .$$

Proof :

For $g \in D(H_1) = D(H_2) = D(K_o)$:

$$(g,H_1g) = (g,K_og) + (g,V_1g)$$

$$\leq (g,K_og) + (g,V_2g) = (g,H_2g) .$$

Hence

$$\rho^{(1)}(f_1,\ldots,f_{n-1}) \leq \rho^{(2)}(f_1,\ldots,f_{n-1})$$

and $$\mu_n^{(1)} \leq \mu_n^{(2)} .$$

Let $m = \dim E^{(2)}_{(-\infty,0)}\mathcal{H}$. Then by Proposition 9.6, $\mu_n^{(1)} \leq \mu_n^{(2)} < 0$ for each $n \leq m$, i.e. H_1 has at least m negative eigenvalues, i.e. $\dim E^{(1)}_{(-\infty,0)}\mathcal{H} \geq m$.∎

LEMMA 9.8 : Let V be as in Example 9.4 and $\varepsilon > 0$. There exists $\alpha = \alpha(\varepsilon) < \infty$ such that

$$|(f,Vf)| \leq \varepsilon(f,K_of) + \alpha\|f\|^2$$

for all $f \in D(K_o)$.

Proof :

$W \equiv \varepsilon^{-1} V$ is K_o-bounded with K_o-bound 0. By Proposition 4.10, and its applications at the end of Chapter 4, $K_o \pm W$ are bounded below, i.e., there exists $-\infty < m \leq 0$ such that

$$(f, (K_o \pm W)f) \geq m\|f\|^2 \qquad \forall f \in D(K_o) ,$$

i.e. $\pm (f,Vf) + \varepsilon(f,K_o f) \geq m\, \varepsilon\|f\|^2$.

If $(f,Vf) \leq 0$, then [take the upper sign]

$$-(f,Vf) = |(f,Vf)| \leq \varepsilon(f,K_o f) + \varepsilon|m|\ \|f\|^2 \quad .$$

If $(f,Vf) \geq 0$ then [take the lower sign]

$$(f,Vf) = |(f,Vf)| \leq \varepsilon(f,K_o f) + \varepsilon|m|\ \|f\|^2 \quad .$$

Thus the lemma holds with $\alpha = \varepsilon|m|$. ∎

LEMMA 9.9 : Let V be as in Example 9.4 and $H_\gamma = K_o + \gamma V$ with $0 \leq \gamma \leq 1$. Let $\mu_n(\gamma)$ be defined as before, with $H = H_\gamma$. Then the functions $\gamma \mapsto \mu_n(\gamma)$ are continuous.

Proof :

(i) For $g \in D(H)$, define

$$t_g(\gamma) = \min\, \{(g,H_\gamma g),0\} \quad .$$

Since $\mu_n \leq 0$, we have

$$\mu_n(\gamma) = \max_{f_1,\ldots,f_{n-1}} \{ \inf_{\substack{g\in D(H),\|g\|=1 \\ g\perp f_k,\forall k=1,\ldots,n-1}} t_g(\gamma)\} \quad .$$

(ii) Assume $\|g\| = 1$ and $(g,H_\gamma g) < 0$ for some $\gamma \in [0,1]$, i.e.

$$(g,K_o g) + \gamma(g,Vg) < 0 \quad .$$

Then

$$(g,K_0 g) < -\gamma(g,Vg) \le \gamma|(g,Vg)| \ .$$

By Lemma 9.8, there exists $\alpha < \infty$ such that

$$|(g,Vg)| \le \frac{1}{2}(g,K_0 g) + \alpha \quad \text{for all} \quad g \in D(H), \ \ \|g\| = 1 \ .$$

Hence, if $(g,H_\gamma g) < 0$, then

$$|(g,Vg)| \le \frac{1}{2}\gamma|(g,Vg)| + \alpha \le \frac{1}{2}|(g,Vg)| + \alpha$$

i.e. $$|(g,Vg)| \le 2\alpha.$$

(iii) If $t_g(\gamma) = 0$ for all $\gamma \in [0,1]$, then

$$|t_g(\gamma) - t_g(\delta)| = 0 \ .$$

If $t_g(\gamma) < 0$ for some $\gamma \in [0,1]$, then by (ii)

$$|(g,Vg)| \le 2\alpha \ .$$

We fix $\gamma,\delta \in [0,1]$ and consider various cases :

(α) It $t_g(\gamma) < 0$ and $t_g(\delta) < 0$:

$$|t_g(\gamma) - t_g(\delta)| = |(g,H_\gamma g) - (g,H_\delta g)|$$

$$= |(\gamma-\delta)(g,Vg)| \le |\gamma-\delta|\ |(g,Vg)|$$

$$\le 2\alpha|\gamma-\delta| .$$

(β) It $t_g(\gamma) < 0$ and $t_g(\delta) = 0$, then $(g,H_\gamma g) < 0$ and $(g,H_\delta g) \ge 0$. Hence

$$|t_g(\gamma) - t_g(\delta)| = |(g,H_\gamma g)| \le |(g,H_\gamma g) - (g,H_\delta g)|$$

$$\le 2\alpha|\gamma-\delta| \quad \text{as above .}$$

It follows that in all cases

$$|t_g(\gamma) - t_g(\delta)| \le 2\alpha|\gamma-\delta| \quad , \text{ for all } \quad g \in D(H)$$

with $\|g\| = 1$.

In view of (i), this implies

$$\Big| \inf_{\substack{g\in D(H),\|g\|=1 \\ g\perp f_k,\forall k=1,\dots,n-1}} t_g(\gamma) - \inf_{\substack{g\in D(H),\|g\|=1 \\ g\perp f_k,\forall k=1,\dots,n-1}} t_g(\delta)\Big| \le 2\alpha|\gamma-\delta|$$

and hence

$$|\mu_n(\gamma) - \mu_n(\delta)| \le 2\alpha|\gamma-\delta| ,$$

i.e. μ_n is continuous. ∎

<u>LEMMA 9.10</u> : Let V be as in Example 9.4 and $V(x) \le 0$ almost everywhere. Let $H_\gamma = K_o + \gamma V$ with $0 \le \gamma \le 1$.

(a) If $\mu_n(\gamma_0) < 0$ and $\gamma > \gamma_0$, then $\mu_n(\gamma) < \mu_n(\gamma_0)$.

(b) If $\lambda < 0$, the number of eigenvalues of $H_1 = K_0 + V$ less than or equal to λ (counting multiplicities) is equal to the number of values of γ in $(0,1]$ for which $-\frac{1}{\gamma}$ is an eigenvalue of $W_\lambda \equiv B(K_0-\lambda)^{-1} A$ (counting multiplicities).

Proof :

(i) By Proposition 9.5(b), if $\rho \neq 0$ is an eigenvalue of H_γ and $M_{\{\rho\}}$ the corresponding eigensubspace then $\dim M_{\{\rho\}} = \dim N(\rho)$, where

$$N(\rho) = \{f \mid (I + \gamma W_\rho)f = 0\} .$$

In particular $-\gamma^{-1}$ is an eigenvalue of W_ρ. Since W_ρ is compact, there exists a neighbourhood U of $-\frac{1}{\gamma}$ such that W_ρ has no eigenvalue other than $-\frac{1}{\gamma}$ (see Lemma 9.11). Hence there exists a neighbourhood V of γ such that the only point δ in V for which ρ is an eigenvalue of H_δ is $\delta = \gamma$.

(ii) Since $V \leq 0$, we have $(g,H_\gamma g) \leq (g,H_{\gamma_0} g)$ whenever $\gamma > \gamma_0$. Hence $\mu_n(\gamma) \leq \mu_n(\gamma_0)$ for $\gamma > \gamma_0$, i.e. μ_n is non-increasing.

If $\mu_n(\gamma_0) < 0$, it coincides with an eigenvalue of H_{γ_0}. If $\mu_n(\gamma) = \mu_n(\gamma_0)$ for some $\gamma > \gamma_0$, then $\mu_n(\delta) = \mu_n(\gamma_0)$ for all $\delta \in [\gamma_0,\gamma]$, which contradicts (i). Hence $\mu_n(\gamma) < \mu_n(\gamma_0)$, proving (a).

(iii) Let $n(\lambda)$ be the number of eigenvalues of $H_1 = K_0 + V$ less than or equal to λ $(\lambda < 0)$. Then, by Proposition 9.6, part (a) of this lemma, and Lemma 9.9 :

$$\begin{aligned} n(\lambda) &= \max \{n \mid \mu_n(1) \leq \lambda\} \\ &= \{\text{number of values } k \mid \mu_k(\gamma) = \lambda \text{ for some } \gamma \in (0,1]\} \end{aligned}$$

$= \{\text{number of } \gamma \in (0,1] \mid \gamma \text{ is eigenvalue of } H_\gamma\}.$
$= \{\text{number of } \gamma \in (0,1] \mid -\frac{1}{\gamma} \text{ is eigenvalue of } W_\lambda\}.$

This proves (b). ∎

As a last preliminary result, we relate the eigenvalues of a general Hilbert-Schmidt operator to its Hilbert-Schmidt norm.

LEMMA 9.11 : Let $A \in \mathcal{B}_\infty$ and $\varepsilon > 0$. Then the number of eigenvalues α_k of A (multiplicities counted) such that $|\alpha_k|^2 > \varepsilon$ is finite. If $A \in \mathcal{B}_2$ then

$$\sum_k |\alpha_k|^2 \leq \|A\|_{HS}^2 ,$$

where the sum is over *all* non-zero eigenvalues, multiplicities counted.

Proof :

(i) Let $A \in \mathcal{B}_\infty$. Let $\{f_n\}$ be an infinite sequence of linearly independent eigenvectors of A, $\{\alpha_k\}$ the corresponding sequence of eigenvalues. Choose an orthonormal set $\{e_i\}$ such that e_n is a linear combination of $f_1,\ldots,f_n$, which can be done by the Schmidt orthogonalization procedure. Thus

$$e_n = \sum_{i=1}^{n} c_{ni} f_i ,$$

and f_k is a linear combinaison of $e_1,\ldots,e_k$. Now

$$Ae_n = \sum_{i=1}^{n} c_{ni}\alpha_i f_i = \sum_{i=1}^{n-1} c_{ni}(\alpha_i-\alpha_n) f_i + \alpha_n e_n$$

$$= \sum_{k=1}^{n-1} d_{nk} e_k + \alpha_n e_n \quad (\text{for some } d_{nk})$$

and

$$\|A\, e_n\|^2 = \sum_{k=1}^{n-1} |d_{nk}|^2 + |\alpha_n|^2 . \tag{9.8}$$

Since $\text{w-}\lim_{n\to\infty} e_n = 0$ and $A \in \mathcal{B}_\infty$, $\lim_{n\to\infty} \|A\, e_n\|^2 = 0$, by Lemma 4.14. In particular $\lim_{n\to\infty} |\alpha_n|^2 = 0$, proving the first assertion.

(ii) Let $A \in \mathcal{B}_2$. Let $\{\alpha_k\}$ be an enumeration of all non-zero eigenvalues of A (multiplicities counted), let $\{f_n\}$ be the corresponding sequence of eigenvectors and construct $\{e_n\}$ as above. Then, by (9.8)

$$\|A\|_{HS}^2 \geq \sum_n \|A\, e_n\|^2 \geq \sum_n |\alpha_n|^2 ,$$

which proves the second assertion. ∎

We now give an upper bound for the number n(0) of negative eigenvalues of H for the case where $\mathcal{H} = L^2(\mathbb{R}^3)$, i.e. n = 3.

<u>PROPOSITION 9.12</u> : In $L^2(\mathbb{R}^3)$, let $H = K_0 + V$, where $V(x) \equiv A(x)\, B(x) = (1 + |x|)^{-\nu} [B_1(x) + B_2(x)]$, $\nu > \frac{3}{2}$, $|B_1(x)| \leq c(1 + |x|)^{-\frac{1}{2}-\varepsilon}$ $(\varepsilon > 0)$ and $B_2(.) \in L^2(\mathbb{R}^3)$. Then the number n_0 of strictly negative eigenvalues of H, multiplicities counted, is finite and bounded by

$$n_0 \leq \|W_o\|_{HS}^2 = \iint dx\, dy\, |B(x)|^2\, |x-y|^{-2}\, (1 + |y|)^{-2\nu}. \tag{9.9}$$

<u>Remark</u> : In Example 9.4, B_2 was required to satisfy a somewhat stronger condition than in Proposition 9.12. It can be shown that all preceding results are true if $B_2(.) \in L^2(\mathbb{R}^3)$, and we use this fact freely.

<u>Proof</u> : Define $B'(x) = -|B(x)|$, $V'(x) = A(x)\, B'(x) = -|V(x)|$, $W_z' = B'(K_0 - z)^{-1} A$ and $H' = K_0 + V'$. Clearly $V'(x) \leq 0$ and $V'(x) \leq V(x)$.

By Corollary 9.7 :

$$n_0 \leq n_0' ,$$

where n_0' is the number of negative eigenvalues of H'.

(ii) By Lemma 9.10(b), for $\lambda < 0$:

$$n'(\lambda) = \{\text{number of } \gamma \in (0,1] \mid -\frac{1}{\gamma} \text{ is eigenvalue of } W_\lambda'\}$$

$$= \{\text{number of eigenvalues of } W_\lambda' \text{ in } (-\infty,-1]\}.$$

We denote by $\{\alpha_k\}$ the non-zero eigenvalues of W_λ'. Then by Lemma 9.11,

$$n'(\lambda) = \sum_{\substack{k \\ \alpha_k<-1}} 1 \leq \sum_{\substack{k \\ \alpha_k<-1}} |\alpha_k|^2 \leq \sum_k |\alpha_k|^2 \leq \|W_\lambda'\|_{HS}^2 , \tag{9.10}$$

so that

$$n'(\lambda) \leq \iint dx\, dy\, |B'(x)|^2 \frac{e^{-\sqrt{|\lambda|}|x-y|}}{|x-y|^2} |A(y)|^2$$

$$\leq \iint dx\, dy\, |B(x)|^2 |x-y|^{-2} |A(y)|^2 = \|W_0\|_{HS}^2 .$$

Hence

$$n_0 \leq n_0' = \lim_{\lambda\to-0} n'(\lambda) \leq \|W_0\|_{HS}^2 \quad \blacksquare$$

Remarks :

(a) It can be shown that $\|W_o\|^2_{HS}$ is also an upper bound for all non-positive eigenvalues, i.e. $\dim E_{(-\infty,0]}\mathcal{H} \le \|W_o\|^2_{HS}$, [SI].

(b) The result of Proposition 9.12 is essentially the so-called *Birman-Schwinger bound*. In fact these authors used a different factorization of V : ([13],[34]) :

$$V = CD \quad ,$$

with C is the multiplication by $|V(x)|^{\frac{1}{2}}$ and D is the multiplication by $|V(x)|^{\frac{1}{2}}$ sign V(x) to obtain similarly

$$n_0 \le \|C(K_0)^{-1} D\|^2_{HS} = \iint dx\, dy\, |V(x)||V(y)||x-y|^{-2} \; .$$

[RS, vol. IV].

In n dimensions (n > 3) the Hilbert-Schmidt norm is not so useful to obtain bounds on $n(\lambda)$. In this case one may use the following expression (sometimes called *the weak* B_p*-norm*) :

$$|||A|||_p = \sup_{n\ge 1} n^{\frac{1}{p}} s_n(A) \quad , \tag{9.11}$$

where $p \ge 1$, A is a compact operator and $s_n(A) = [\lambda_n(A^*A)]^{\frac{1}{2}}$, where $\{\lambda_n(A^*A)\}$ are the non-zero eigenvalues of the positive compact operators (A^*A) arranged in decreasing order (counting multiplicities).

Let us write $V(x) = V_+(x) - V_-(x)$, with $V_\pm(x) \ge 0$. Then, by Corollary 9.7, $n(\lambda) \le n_-(\lambda)$ where $n_-(\lambda)$ is the number of eigenvalues less than or equal to λ ($\lambda < 0$) of $H_- = K_0 - V_-$. As in the proof of Proposition 9.12, one will have $n_-(\lambda)$ = the number of eigenvalues of $V_-^{1/2}(K_0-\lambda)^{-1} V_-^{1/2}$ in $[1,\infty)$.

If we set $A_- = (K_0-\lambda)^{-1/2}\, V_-^{1/2}$, then

$$n_-(\lambda) = \max_{n\geq 1} \{n \mid s_n(A_-) \geq 1\} \leq \|A_-\|_p^p \tag{9.12}$$

for any $p \geq 1$, provided $A_- \in \mathcal{B}_\infty$. Therefore, in order to obtain a bound for n_0, it suffices to find conditions on V_- such that $A_- \in \mathcal{B}_\infty$ and to estimate $|||A_-|||_p$ in terms of V_-. We indicate below a way of doing this, following a paper by (Cwickel, [15]).

Since $k \mapsto (k^2-\lambda)^{-1/2} \in L^p(\mathbb{R}^n)$ for any $p > n$ $(\lambda < 0)$, it follows by Proposition 2.8 that $A_- \in \mathcal{B}_\infty$ provided that $V_-(.) \in L^q(\mathbb{R}^n)$ for some q satisfying $q > \frac{n}{2}$. Proposition 2.8 also implies that $\|V_-^{1/2}(K_0-\lambda)^{-1}\, V_-^{1/2}\| \to 0$ as $\lambda \to -\infty$. This means that V_- is a small form pertubation of K_0, and the sum $K_0 + V$ can then be defined by using the associated quadratic forms. By using for example Proposition 9.13 below, one sees that this is still possible if $q = \frac{n}{2}$. We do not go into the details of the definition of H and refer to (Faris, [21], § 4 and § 5). We note that, in deriving the results of Lemma 9.10, we used only the forms (g,Vg) and (g,H_0g) associated with V and H_0, so that Lemma 9.10 applies to H_- if $V_- \in L^q(\mathbb{R}^n)$, $q \geq \frac{n}{2}$. (For Proposition 9.5 when $\lambda < 0$, it suffices to notice that $D(|H|^{1/2}) = D(K_0^{1/2}) \subseteq D(V_-^{1/2})$.).

In Cwickel's paper one of the L^p-norms in Proposition 2.8 is replaced by a "weak L^p-norm" defined as

$$\|f\|_{p,w} = \sup_{\alpha>0} [\alpha^p \mu\{x \in \mathbb{R}^n \mid |f(x)| > \alpha\}]^{\frac{1}{p}}, \tag{9.13}$$

where $\mu(\Delta)$ denotes the Lebesgue measure of Δ. Then

<u>PROPOSITION 9.13</u> (Cwickel) : Let $2 < p < \infty$, let $\phi \in L^p(\mathbb{R}^n)$ and $\psi \in L_w^p(\mathbb{R}^n)$, i.e. $\|\psi\|_{p,w} < \infty$. Then the operator $\phi(Q)\, \psi(P)$ is compact, and there is a constant $c_{n,p}$ such that

$$\|\phi(Q)\ \psi(P)\|_p \le c_{n,p}\ \|\phi_p\|\|\psi\|_{p,w}\ . \tag{9.14}$$

Before giving the proof of this theorem, we apply it to the problem which interests us here. In our application $\phi(x) = [V_-(x)]^{1/2}$ and $\psi(k) = (k^2-\lambda)^{-1/2}$ $(\lambda < 0)$; one notices that $\psi \in L^p_w(\mathbb{R}^n)$ for all $p \in [n,\infty)$, and that

$$\|\psi\|^p_{p,w} = \begin{cases} (n^{-1}\ a_n) & \text{if} \quad p = n \\ (n^{-1}\ a_n)(\frac{p-n}{p|\lambda|})^{\frac{1}{2}(p-n)}\ (\frac{n}{p})^{\frac{n}{2}} & \text{if} \quad p > n\ , \end{cases}$$

where a_n denotes the surface area of the unit sphere S^{n-1} in $\mathbb{R}^n$. As $\lambda \to -0$, $\|\psi\|^p_{p,w} \to \infty$ if $p > n$. However, for $p = n$, $\|\psi\|_{p,w}$ is independent of λ. Therefore Proposition 9.13 gives a bound on $n_-(\lambda)$ which does not depend on λ :

PROPOSITION 9.14 : Suppose $V = V_+ - V_-$ is such that $V_- \in L^{n/2}(\mathbb{R}^n)$, $n \ge 3$. Then the number n_0 of negative eigenvalues of $H = K_0 + V$ (counting multiplicities) is bounded by

$$n_0 \le C_n\ \|V_-\|^{\frac{n}{2}}_{\frac{n}{2}}\ , \tag{9.15}$$

for some constant $C_n < \infty$.

Proof of Proposition 9.13 :

Without loss of generality, we may assume that $\|\phi\|_p = \|\psi\|_{p,w} = 1$.

(i) For $\alpha \in \mathbb{R}$, write $\mathbb{R}^n \times \mathbb{R}^n = \Delta^+_\alpha \cup \Delta^-_\alpha$ where

$$\Delta^+_\alpha = \{(x,k) \in \mathbb{R}^n \times \mathbb{R}^n \mid |\phi(x)\ \psi(k)| \ge 2^\alpha\}.$$

$$\Delta^-_\alpha = \{(x,k) \in \mathbb{R}^n \times \mathbb{R}^n \mid |\phi(x)\ \psi(k)| < 2^\alpha\}.$$

Notice that by (9.13), the Lebesgue measure of Δ_α^+ is

$$\mu(\Delta_\alpha^+) = \int dx\ \mu\{k \in \mathbb{R}^n \mid |\psi(k)| \geq 2^\alpha\ |\phi(x)|^{-1}\}$$

$$\leq \int dx\ |2^{-\alpha}\ \phi(x)|^p\ \|\psi\|_{p,w}^p$$

$$= 2^{-\alpha p} \|\phi\|_p^p\ \|\psi\|_{p,w}^p = 2^{-\alpha p}\ .$$

We write $\phi(Q)\ \psi(P) = T_m^+ + T_m^-$, where m is an integer and

$$(T_m^\pm\ f)(x) = (2\pi)^{-\frac{n}{2}}\ \phi(x) \int e^{ik.x}\ \chi_{\Delta_m^\pm}\ (x,k)\ \psi(k)\ \tilde{f}(k)\ dk\ .$$

(χ_Δ denotes the characteristic function of the set Δ).

(ii) Bound on $\|T_m^+\|_{HS}$. One obtains as in Proposition 2.7 that

$$\|T_m^+\|_{HS}^2 = (2\pi)^{-n} \iint_{\Delta_m^+} |\phi(x)|^2\ |\psi(k)|^2\ dx\ dk$$

$$= (2\pi)^{-n}\ 2^{2m}\ \mu(\Delta_m^+) + (2\pi)^{-n} \int_m^\infty \mu(\Delta_\alpha^+)\ d\ (2^{2\alpha})$$

$$\leq (2\pi)^{-n}\ 2^{m(2-p)} + (2\pi)^{-n}\ \frac{2}{p-2}\ 2^{m(2-p)}$$

$$= (2\pi)^{-n}\ \frac{p}{p-2}\ 2^{m(2-p)}\ . \tag{9.16}$$

(iii) We now estimate $\|T_m^-\|$. For each $\ell = 0,\pm1,\pm2,\ldots$, we introduce the following subsets of $\mathbb{R}^n$:

$$X_\ell = \{x \in \mathbb{R}^n \mid 2^{\ell-1} < |\phi(x)| \leq 2^\ell\},$$

$$K_\ell = \{k \in \mathbb{R}^n \mid 2^{\ell-1} < |\psi(k)| \leq 2^\ell\}.$$

For $f,g \in L^2$, we define f_ℓ and g_ℓ by

$$f_\ell(x) = 2^{-\ell} \phi(x) f(x) \chi_{X_\ell}(x) ,$$

$$\tilde{g}_\ell(k) = 2^{-\ell} \psi(k) \tilde{g}(k) \chi_{K_\ell}(k) .$$

Then, by the Schwarz inequality for the integral and the sum :

$$|(f,T_m^- g)| \le |\sum_{\ell+s\le m+1}\sum 2^{\ell+s} (2\pi)^{-\frac{n}{2}} \iint e^{ik.x}\, \overline{f_\ell(x)}\, \tilde{g}_s(k)\, dx\, dk|$$

$$\le \sum_{\ell+s\le m+1}\sum 2^{\ell+s} |\int \overline{\tilde{f}_\ell(k)}\, \tilde{g}_s(k)\, dk|$$

$$\le \sum_{r=-\infty}^{m+1} 2^r \sum_{\ell=-\infty}^{\infty} \|f_\ell\|\, \|g_{r-\ell}\|$$

$$\le \sum_{r=-\infty}^{m+1} 2^r \left(\sum_\ell \|f_\ell\|^2\right)^{\frac{1}{2}} \left(\sum_\ell \|\tilde{g}_{r-\ell}\|^2\right)^{\frac{1}{2}} .$$

Now, since $|f_\ell(x)| \le |f(x)|\, \chi_{X_\ell}(x)$ and $|\tilde{g}_s(k)| \le |\tilde{g}(k)|\, \chi_{K_s}(k)$, we obtain

$$|(f, T_m^- g)| \le 2^{m+2} \|f\|\, \|g\| ,$$

which implies that

$$\|T_m^-\| \le 2^{m+2} . \tag{9.17}$$

(iv) (9.17) implies that

$$\lim_{m\to-\infty} \|\phi(Q)\, \psi(P) - T_m^+\| = 0 ,$$

hence $\phi(Q)\ \psi(P) \in \mathcal{B}_\infty$ as the uniform limit of a sequence of Hilbert-Schmidt operators (see Lemma 2.4). We now use the following two inequalities which will proved in (v) below :

$$j^{\frac{1}{2}}\ s_j(A) \leq \|A\|_{HS} \tag{9.18}$$

$$s_j(B+C) \leq s_j(B) + \|C\| \quad (B,C \in \mathcal{B}_\infty). \tag{9.19}$$

We then get that, for any $m = 0,\pm1,\pm2,\ldots$:

$$s_j(\phi(P)\ \psi(Q)) = s_j(T_m^+ + T_m^-)$$

$$\leq j^{-\frac{1}{2}}\ \|T_m^+\|_{HS} + \|T_m^-\| \leq c_{n,p}\ j^{-\frac{1}{2}}\ 2^{m(1-\frac{p}{2})} + 2^{m+2} .$$

If we choose $m \leq 0$ such that $2^{m-1} < j^{-\frac{1}{p}} \leq 2^m$, then, since $p > 2$:

$$s_n(\phi(P)\ \psi(Q)) \leq c'_{n,p}\ j^{-\frac{1}{p}} . \tag{9.20}$$

In view of the definition (9.11), this implies that $\|\phi(Q)\ \psi(P)\|_p \leq c'_{n,p}$, which proves the second assertion of the Proposition.

(v) It remains to prove (9.18) and (9.19). We have

$$\|A\|^2_{HS} = \sum_{i=1}^{\infty} s_i^2 \geq \sum_{i=1}^{m} s_i^2 \geq \sum_{i=1}^{m} s_m^2 = m\ s_m^2 ,$$

which proves (9.18). (9.19) is obtained by using the mini-max principle for calculating the eigenvalues λ_m of A^*A (see Proposition 9.6) :

$$\lambda_m = \min_{f_1,\ldots,f_{m-1}} \operatorname*{Sup}_{\substack{\|g\|=1 \\ g\perp f_k}} (g, A^*A\, g).$$

This leads to

$$s_m(A) = \min_{f_1,\ldots,f_{m-1}} \operatorname*{Sup}_{\substack{\|g\|=1 \\ g\perp f_k}} \|Ag\|. \tag{9.21}$$

Since $\|Bg + Cg\| \leq \|Bg\| + \|Cg\|$, (9.21) implies (9.19). ∎

10 Periodic potentials

The purpose of this chapter is twofold. The ultimate goal is to give a condition on a potential $V(x)$ on $\mathbb{R}^n$ which guarantees that with $H = K_0 + V$, $\sigma_p(H)$ contains no points in $(0,\infty)$. The approach we take to this is to reduce it to showing that, for a broad class of periodic potentials $W(x)$ on $\mathbb{R}^n$, $H = K_0 + W$ has no point spectrum. Indeed, we show $K_0 + W$ has purely absolutely continuous spectrum, under certain mild restriction on a *periodic* potential W. Interestingly enough, this involves the study of families of operators with purely discrete spectra, including some results on perturbation of operators with discrete spectra.

A function $W : \mathbb{R}^n \to \mathbb{R}$ is said to be *periodic* if there are n linearly independent vectors $a_1,\ldots,a_n$ in $\mathbb{R}^n$ such that $W(x+a_i) = W(x)$ for all $x \in \mathbb{R}^n$ and $i = 1,\ldots,n$. For the sake of simplicity we assume that the basic cell determined by $a_1,\ldots,a_n$ is a cube with side L, i.e. that

$$a_i . a_k = L^2 \; \delta_{ik} \quad . \tag{10.1}$$

We denote by C the cube $C = \{ \sum_{i=1}^{n} \alpha_i a_i \mid 0 \le \alpha_i < 1\}$.

We first give a result on self-adjointness of Schrödinger operators with periodic potentials.

PROPOSITION 10.1 : Let $W : \mathbb{R}^n \to \mathbb{R}$ be periodic and such that

$$\|W\|_{s,C} \equiv \left(\int_C |W(x)|^s \, dx\right)^{\frac{1}{s}} < \infty \; , \tag{10.2}$$

where s satisfies $s \ge 2$ and $s > \frac{n}{2}$. Then the multiplication operator W by the function $W(x)$ is K_0-bounded with K_0-bound 0. In particular, $H = K_0 + W$

is self-adjoint on $D(K_o)$.

Proof : Let q be such that $q^{-1} = \frac{1}{2} - s^{-1}$ and q' such that $(q')^{-1} = 1 - q^{-1} = \frac{1}{2} + s^{-1}$. Let $f \in D(K_o)$ and $\mu > 0$. Then, by the Hausdorff-Young inequality (2.11) and the Hölder inequality (2.10) :

$$\|f\|_q^2 \le \|\tilde{f}\|_{q'}^2 \le \|(\mu+k^2)^{-1}\|_s^2 \ \|(\mu+k^2)\ \tilde{f}\|_2^2 . \tag{10.3}$$

Given $\varepsilon > 0$, we choose μ so large that $\|(\mu+k^2)^{-1}\|_s^2 < \frac{\varepsilon}{2}$. (10.3) then implies that

$$\|f\|_q^2 \le \varepsilon\ \|K_o f\|_2^2 + \varepsilon\ \mu^2\ \|f\|_2^2 . \tag{10.4}$$

We shall denote by $m = (m_1,\ldots,m_n)$ the points in $\mathbb{Z}^n$. Let C_m be the cube

$$C_m = \{Lm + \sum_{i=1}^{n} \alpha_i\ a_i \mid 0 \le \alpha_i < 1\}. \tag{10.5}$$

Then $\mathbb{R}^n = \bigcup_m C_m$. Furthermore, let $\phi \in C_o^\infty(\mathbb{R}^n)$ be such that $0 \le \phi(x) \le 1$, $\phi(x) = 1$ for $x \in C$ and define ϕ_m by $\phi_m(x) = \phi(x-Lm)$. Notice that $\phi_m(x) = 1$ for $x \in C_m$. Let B be a ball in $\mathbb{R}^n$ containing supp ϕ in its interior, and notice that supp $\phi_m \subset B_m \equiv B + Lm$. Using (6.28), the fact that there is an integer N such that each point $x \in \mathbb{R}^n$ is contained in at most N of the balls $\{B_m\}_{m \in \mathbb{Z}^n}$, and the fact that $|P|$ is K_o-bounded with K_o-bound 0, we get that

$$\sum_m \|K_o\phi_m(Q)f\|_2^2 \le 3\sum_m \|\phi_m K_o f\|_2^2$$

$$+ 3\sum_m \|(\Delta\phi_m)f\|_2^2 + 6n \sum_{j=1}^{n}\sum_m \|\frac{\partial\phi}{\partial x_j} P_j f\|_2^2$$

$$\le 3N\|K_o f\|_2^2 + 3N\|\Delta\phi\|_\infty^2 \|f\|_2^2$$

$$+ 6nN \| |\nabla\phi| \|_\infty^2 \{\|K_o f\|_2^2 + b\|f\|_2^2\} . \tag{10.6}$$

Together with (10.4), this now implies that

$$\|Wf\|_2^2 = \sum_m \|Wf\|_{2,C_m}^2 \le \sum_m \|W\|_{s,C_m}^2 \|f\|_{q,C_m}^2$$

$$\le \|W\|_{s,C}^2 \sum_m \|\phi_m(Q)f\|_q^2 \tag{10.7}$$

$$\le 3\varepsilon N(1+2n \| |\nabla\phi| \|_\infty^2) \|W\|_{s,C}^2 \|K_o f\|^2 + c_n(\varepsilon,\phi) \|W\|_{s,C}^2 \|f\|_2^2.$$

Since ε is arbitrary, this shows that W is K_o-bounded with K_o-bound 0. Hence $K_o + W$ is self-adjoint by Proposition 3.6. ■

To study the spectral properties of $H = K_o + W$, consider the following unitary representation in $L^2(\mathbb{R}^n)$ of the additive group $\mathbb{Z}^n$:

$$[U(m)f](x) = f(x - \sum_{i=1}^{n} m_i a_i) = f(x - Lm) \quad , \quad (m \in \mathbb{Z}^n), \tag{10.8}$$

where we have written $\sum_i m_i a_i = Lm$ by assuming that the directions of the vectors a_i coincide with the Cartesian coordinate system in $\mathbb{R}^n$. Clearly both K_o and W, hence also $H = K_o + W$, commute with U(m). It is therefore useful to represent the Hilbert space $L^2(\mathbb{R}^n)$ in a way that takes this special structure into account.

For this, we divide $\mathbb{R}^n$ into a second union of cubes : $\mathbb{R}^n = \bigcup_m \Gamma_m$, where

$$\Gamma_m = \{k \in \mathbb{R}^n \mid k = \frac{2\pi}{L} m + \sum_{i=1}^{n} \alpha_i e_i \quad , \quad 0 \leq \alpha_i < 1\} \tag{10.9}$$

and $e_1,\ldots,e_n$ are n vectors satisfying

$$e_i \cdot a_j = 2\pi \, \delta_{ij} \; .$$

It is seen that each Γ_m is a cube with side $\ell \equiv \frac{2\pi}{L}$. The cube Γ_o (i.e. $m = (0,0,\ldots,0)$) will be denoted simply by Γ.

We now consider the Hilbert space $G = L^2(\Gamma\,;\,\ell^2(\mathbb{Z}^n))$ of all measurable square integrable function defined on Γ with values in the sequence space $\ell^2(\mathbb{Z}^n)$. We write $f(k)_m$ for the component m ($m \in \mathbb{Z}^n$) of f at the point $k \in \Gamma$. Thus

$$\|f\|_G^2 = \int_\Gamma dk \sum_{m \in \mathbb{Z}^n} |f(k)_m|^2 \; . \tag{10.10}$$

The equation

$$(Uf)(k)_m = \tilde{f}(k + \ell m) \tag{10.11}$$

defines a unitary operator from $L^2(\mathbb{R}^n)$ onto G, as is easily seen from Plancherel's theorem for the Fourier transformation. The inverse of U is as follows : If $g = \{g(k)\} \in G$, then

$$(FU^{-1}g)(\xi) = g(k)_m \, ,$$

where $m \in \mathbb{Z}^m$ and $k \in \Gamma$ are determined by the relation $k + \ell m = \xi$.

For each $k \in \Gamma$, let us define an operator $H_0(k)$ in $\ell^2(\mathbb{Z}^n)$ by

$$(H_0(k)\ g)_m = (k + \ell m)^2\ g_m \quad , \quad g \in \ell^2(\mathbb{Z}^n)\ . \tag{10.12}$$

$$D(H_0(k)) = D_0 = \{g \in \ell^2(\mathbb{Z}^n) \mid \sum_{m \in \mathbb{Z}^n} |m^2 g_m|^2 < \infty\}.$$

We then have

LEMMA 10.2 :

(i) For each $k \in \Gamma$, $H_0(k)$ is self-adjoint.

(ii) For each $\mu \in \rho(H_0(k))$, the resolvent $(H_0(k)-\mu)^{-1}$ of $H_0(k)$ is a compact operator.

(iii) One has

$$(\mathcal{U}H_0 f)(k) = H_0(k)(\mathcal{U}f)(k) \tag{10.13}$$

in the following sense : a vector f in $L^2(\mathbb{R}^n)$ is in $D(H_0)$ if and only if $(\mathcal{U}f)(k) \in D_0$ a.e. and

$$\int_\Gamma dk\ \|H_0(k)(\mathcal{U}f)(k)\|^2_{\ell^2(\mathbb{Z}^n)} < \infty \ ,$$

and the vector $\mathcal{U}H_0 f$ is then given by (10.13) for almost all $k \in \Gamma$.

Proof :

(i) $H_0(k)$ is a multiplication operator by a real-valued function in an L^2-space (with discrete measure), hence it is self-adjoint by Proposition 3.4.

(iii) This follows immediately from the definitions (10.11) and (10.12) and the fact that $(\widetilde{H_0 f})(k) = k^2\tilde{f}(k)$.

(ii) $[H_0(k) - \mu]^{-1}$ is the multiplication operator by $\psi(m) = [(k + \ell m)^2 - \mu]^{-1}$. Let D_M be the multiplication operator by $\psi(m)\chi_M(m)$, where $\chi_M(m) = 1$ if $m^2 \leq M$ and $\chi_M(m) = 0$ if $m^2 > M$. D_M is a compact operator since it acts non-trivially only in a finite-dimensional subspace of $\ell^2(\mathbb{Z}^n)$. Furthermore

$$\|[H_0(k) - \mu]^{-1} - D_M\| = \sup_{m^2 > M} |(k + \ell m)^2 - \mu|^{-1}$$

tends to 0 as $M \to \infty$, so that $[H_0(k) - \mu]^{-1}$ is compact as the uniform limit of a sequence $\{D_M\}$ of compact operators. ∎

We denote by $\{w_m\}_{m \in \mathbb{Z}^n}$ the Fourier coefficients of the periodic function $W(x)$:

$$w_m = L^{-\frac{n}{2}} \int_C dx\, e^{-i\ell m.x}\, W(x) . \tag{10.14}$$

We may then define the following operator Y in $\ell^2(\mathbb{Z}^n)$:

$$(Yg)_m = L^{-\frac{n}{2}} \sum_{q \in \mathbb{Z}^n} w_q\, g_{m-q} . \tag{10.15}$$

LEMMA 10.3 : Let W satisfy the assumptions of Proposition 10.1. Then :

(i) $D_0 \subseteq D(Y)$ and Y is symmetric on D_0.

(ii) Y is $H_0(k)$-bounded with $H_0(k)$-bound 0.

(iii) For $\mu \in \rho(H_0(k))$, $Y(H_0(k) - \mu)^{-1}$ is compact.

(iv) If $f \in D(H_0)$, then

$$(\mathcal{U}Hf)(k) = [H_0(k) + Y]\,(\mathcal{U}f)(k)\ . \tag{10.16}$$

<u>Proof</u> : To prove (ii) and (iii), we use the analogue of Proposition 2.8 for Fourier series (the proof is essentially identical with that of Proposition 2.8). $Y[H_0(k) - \mu]^{-1}$ is an operator in $\ell^2(\mathbb{Z}^n)$ of the form

$$(Ag)_m = \sum_{q \in \mathbb{Z}^n} w_{m-q}\, b_q\, g_q\ , \tag{10.17}$$

where $b_q = [(k + \ell q)^2 - \mu]^{-1}$. We notice that $\{b_q\} \in L^r(\mathbb{Z}^n)$ for each $r > \frac{n}{2}$.

Since $W \in L^s(C)$ with $s \geq 2$, $s > \frac{n}{2}$, Proposition 2.8 implies that A is compact and that

$$\|A\| \leq (2\pi)^{\frac{n}{2} - \frac{n}{s}}\ \|W\|_{L^s(C)}\ \|\{b_q\}\|_{\ell^s(\mathbb{Z}^n)}\ . \tag{10.18}$$

The norm $\|\{b_q\}\|_s$ depends on μ and converges to zero as $\mu \mapsto +\infty$. Hence, given $\varepsilon > 0$, we may choose μ such that $\|Y[H_0(k) - \mu]^{-1}\| < \varepsilon$. We then have for $f \in D_0$:

$$\|Yf\| \leq \|Y[H_0(k) - \mu]^{-1}\|\ \{\|H_0(k)f\| + \mu\|f\|\}$$

$$\leq \varepsilon\ \|H_0(k)f\| + c\|f\|\ ,$$

which shows that Y is $H_0(k)$-bounded with $H_0(k)$-bound zero.

The symmetry of Y on D_0 follows easily by observing that $\overline{w_{-m}} = w_m$. To prove (10.16), it suffices to show that $(\mathcal{U}Wf)(k) = Y(\mathcal{U}f)(k)$, which can be done by simply calculating the Fourier transform of Wf. ∎

The proof of the spectral properties of $K_0 + W$ for periodic W uses an argument involving analytic functions. Analyticity enters naturally if one observes that $H_0(k)$ is multiplication by $(k + \ell m)^2$ in $\ell^2(\mathbb{Z}^n)$ $(m \in \mathbb{Z}^n)$, which is a holomorphic function of a vector $k \in \Gamma \times i\mathbb{R}^n$. It will suffice to consider the simplest case where only one component of k is allowed to become complex. For this we introduce the following notations : we let

$$\Gamma' = \{p \in \mathbb{R}^{n-1} \mid p = \sum_{i=1}^{n-1} \alpha_i e_i \ , \ 0 \le \alpha_i < 1\}$$

and

$$G = \{(\kappa + i\nu) \mid \kappa \in (0,1) \ , \ \nu \in \mathbb{R}\} .$$

The first n-1 components of $k \in \Gamma$ determine a vector in Γ'. The last component of k is allowed to assume complex values in the domain $G(\ell) = \{\ell z \mid z \in G\}$.

For $p \in \Gamma'$, $z \in G$ we now denote by $H_0(p,z)$ the multiplication operator in $\ell^2(\mathbb{Z}^n)$ by $(p + ze_n + \ell m)^2$.

LEMMA 10.4 :

(i) For each fixed $p \in \Gamma'$, $\{H_0(p,z)\}$ is a self-adjoint holomorphic family of operators of type (A) in the domain G. In other words :

(α) $H_0(p,\bar{z}) = [H_0(p,z)]^*$.

(β) the domain of the operator $H_0(p,z)$ is independent of z : $D(H_0(p,z)) = D_0$ for all $z \in G$.

(γ) if $\zeta \in \rho(H_0(p,z_0))$ for some z_0, then the function $z \mapsto H_0(p,z)\,[H_0(p,z_0) - \zeta]^{-1}$ is holomorphic from G to $\mathcal{B}(\ell^2(\mathbb{Z}^n))$ (i.e. differentiable in the uniform sense).

(ii) For each $p \in \Gamma'$ and $z \in G$, the resolvent of $H_0(p,z)$ is compact.

(iii) If Im $z \neq 0$, then $\lambda = 0$ belongs to the resolvent set $\rho(H_0(p,z))$

of $H_0(p,z)$.

(iv) For each fixed $p \in \Gamma'$ and $z \in G$, there is a sequence $\tau_n \to +\infty$ such that $i\tau_n \in \rho(H_0(p,z))$.

Proof :

(i) is easy to verify, and (ii) is proven in the same way as Lemma 10.2 (ii).

(iii) For $z = \kappa + i\nu \in G$, we have

$$\mathrm{Im}(p + ze_n + \ell m)^2 = 2\ell^2\nu(\kappa + m_n) , \tag{10.19}$$

which is non-zero if $\nu \neq 0$, since $\kappa \in (0,1)$ and $m_n \in \mathbb{Z}$. Hence

$$\|[H_0(p,z)]^{-1}\| = \sup_{m \in \mathbb{Z}^n} \left|\frac{1}{(p+ze_n+\ell m)^2}\right| < \infty ,$$

which shows that $0 \in \rho(H_0(p,z))$. The proof of (iv) is similar. ∎

LEMMA 10.5 : Let W satisfy the assumptions of Proposition 10.1 and define $H(p,z) = H_0(p,z) + Y$. Then

(i) For each fixed $p \in \Gamma'$, $H(p,z)$ is a self-adjoint holomorphic family of operators of type (A). In particular, if $\zeta \in \rho(H(p,z_0))$ for some z_0, then $z \mapsto H(p,z)\,[H(p,z_0) - \zeta]^{-1}$ is holomorphic from G to $\mathcal{B}(\ell^2(\mathbb{Z}^n))$.

(ii) For each $p \in \Gamma'$ and $z \in G$, the resolvent of $H(p,z)$ is compact, and the resolvent set of $H(p,z)$ is non-empty.

Proof :

(i) follows immediately from Lemma 10.4 (i) and 10.3. Notice that

$$H(p,z)[H(p,z_0) - \zeta]^{-1} = [H_0(p,z)+Y][H_0(p,z_0)-\zeta']^{-1}[I-(Y+\zeta'-\zeta)(H(p,z_0)-\zeta)^{-1}].$$

The product of the first two factors is holomorphic in z by Lemma 10.4 (i) if we choose $\zeta' \in \rho(H_0(p,z_0))$, and the third factor is in $\mathcal{B}(\ell^2(\mathbb{Z}^n))$.

(ii) We have

$$H(p,z) - \mu = (I + Y[H_0(p,z) - \mu]^{-1})(H_0(p,z) - \mu) .$$

Consequently

$$[H(p,z) - \mu]^{-1} = [H_0(p,z) - \mu]^{-1} (I + Y[H_0(p,z) - \mu]^{-1})^{-1} . \quad (10.20)$$

We shall show below that

$$\lim_{\lambda\to\infty} \| Y[H_0(p,z) - i\lambda]^{-1} \| = 0 . \quad (10.21)$$

Thus the second factor on the r.h.s. of (10.20) is in $\mathcal{B}(\ell^2(\mathbb{Z}^n))$ if $\mu = i\lambda$ and $\lambda \geq \lambda_0$. By Lemma 10.4 it is possible to choose a number $\lambda \geq \lambda_0$ such that $[H_0(p,z) - i\lambda]^{-1} \in \mathcal{B}_\infty$. This shows that $\rho(H(p,z)) \neq \phi$ and that H(p,z) has compact resolvent (see Lemma 4.12).

The proof of (10.21) is based on the observation that the operator occuring in (10.21) is of the form (10.17), with $b_q = [(p + ze_n + \ell q)^2 - i\lambda]^{-1}$. We notice that, if $z = \kappa + i\nu$:

$$|b_q|^2 = \{[(p + \kappa e_n + \ell q)^2 - \ell^2\nu^2]^2 + 4\nu^2\ell^4(\kappa + q_n - \frac{\lambda}{2\nu\ell^2})^2\}^{-1}$$

$$\leq [(p + \kappa e_n + \ell q)^2 - \ell^2\nu^2]^{-2} . \quad (10.22)$$

This shows that each term in the series for $\|\{b_q\}\|_{\ell^s(\mathbb{Z}^n)}$ converges to zero as $\lambda \to \infty$ and that this series is bounded, uniformly in λ, by a convergent series (since $s > \frac{n}{2}$). Hence $\|\{b_q\}\|_{\ell^s(\mathbb{Z}^n)} \to 0$ as $\lambda \to \infty$, and (10.21) follows from (10.18).

(If $\nu = 0$, (10.22) must be modified in an obvious way. If z is such that $(p + \kappa e_n + \ell q)^2 = \ell^2\nu^2$ for certain $q \in \mathbb{Z}^n$, there is $c > 0$ and $\lambda_0 < \infty$ such that $4\nu^2\ell^4[\kappa + q_n - \frac{\lambda}{2\nu\ell^2}]^2 \geq c$ for all such q and all $\lambda \geq \lambda_0$. For these values of q one may take in (10.22) the majorization by c^{-1}). ∎

We shall also need the analogue of Lemma 10.4 (iii) for the family $\{H(p,z)\}$, namely that $0 \in \rho(H(p,z))$ for almost p and z. This will be shown under a stronger integrability assumption on W(x).

<u>LEMMA 10.6</u> : Assume that $\|W\|_{s,C} < \infty$ for some s satisfying $s \geq 2$, $s > n-2$. Then for each fixed $p \in \Gamma'$, there is a $z \in G$ such that $0 \in \rho(H(p,z))$.

<u>Proof</u> : We use (10.20) with $\mu = 0$. By Lemma 10.4 (iii), $[H_0(p,z)]^{-1} \in \mathcal{B}_\infty$ if Im $z \neq 0$, so that it suffices to show that

$$\lim_{\nu\to\infty} \|Y[H_0(p, \kappa + i\nu)]^{-1}\| = 0 . \tag{10.23}$$

For this, we again use (10.17) and (10.18) with $b_q = [p + (\kappa + i\nu)e_n + \ell q]^{-2}$. It follows that, as in (10.22),

$$\|\{b_q\}\|_s^s = \sum_q \{[(p + \kappa e_n + \ell q)^2 - \ell^2\nu^2]^2 + 4\nu^2\ell^4(\kappa + q_n)^2\}^{-\frac{s}{2}}$$

$$= \sum_{q_n \in \mathbb{Z}} \sum_{m \in \mathbb{Z}^{n-1}} \{[(p + \ell m)^2 + b]^2 + a^2\}^{-\frac{s}{2}} , \tag{10.24}$$

with $a = 2\ell^2\nu\,|q_n + \kappa|$, $b = \ell^2[(q_n + \kappa)^2 - \nu^2]$. The sum over $m \in \mathbb{Z}^{n-1}$ can be carried out by comparison with an integral. We refer to (A.M. Berthier,

[10]) for the detailed estimates and just indicate the result : given $p \in \Gamma'$, $\kappa \in (0,1)$ and $\nu_0 > 0$, there is a number $K = K(p,\kappa,\nu_0) < \infty$ such that for all $\nu > \nu_0$:

$$\sum_{m \in \mathbb{Z}^{n-1}} \{[(p + \ell m)^2 + b]^2 + a^2\}^{-\frac{s}{2}}$$

$$\leq K \int_{\mathbb{R}^{n-1}} \{(x^2 + b)^2 + a^2\}^{-\frac{s}{2}} dx. \qquad (10.25)$$

The integral can be evaluated in spherical polar coordinates, and one obtains for the r.h.s. of (10.25) the expression

$$K|S_{n-2}| \, a^{-s+1} [a^2 + b^2]^{\frac{n-3}{4}} \int_\rho^{\frac{\pi}{2}} du \, (\cos u)^{s - \frac{n}{2} - \frac{1}{2}} [\sin(u-\rho)]^{\frac{n-3}{2}} ,$$

where $\rho = \arc\tan (\frac{b}{a})$. The integral $I(\rho)$ over du in this expression may be estimated as follows :

if $b < 0$, i.e. $\rho \in [-\frac{\pi}{2},0)$: $I(\rho) \leq M < \infty$,

if $b > 0$, i.e. $\rho \in (0,\frac{\pi}{2}]$:

$$I(\rho) \leq M(\tfrac{\pi}{2} - \rho)^{s-1} = M'(\tfrac{\pi}{2} - \arc\tan \tfrac{b}{a})^{s-1}$$

$$\leq 2M' \, [1 + (\tfrac{b}{a})^2]^{\frac{1-s}{2}} .$$

Since $a^2 + b^2 = \ell^4[(q_n + \kappa)^2 + \nu^2]^2$, this leads to

$$\|\{b_q\}\|_s^s \leq c_1 \, \nu^{n-3-s+1} \sum_{|q_n+\kappa|<\nu} |q_n + \kappa|^{-s+1}$$

$$+ c_2 \sum_{|q_n+\kappa|\geq\nu} [(q_n + \kappa)^2 + \nu^2]^{-\frac{1}{2}-\theta} ,$$

with $\theta = s - \frac{n}{2} > 0$. As $\nu \to \infty$, the second sum converges to zero since $\theta > 0$, whereas the first sum is $O(\nu^{n-s-2} \log \nu)$, since $s \geq 2$, hence it converges to zero by the hypothesis that $s > n-2$. This proves (10.23). ∎

The next Lemma gives a strenghtening of the result of Lemma 10.6 :

LEMMA 10.7 : Assume the hypotheses of Lemma 10.6. Then, for each fixed $p \in \Gamma'$ and each compact subset G_0 of G, one has $0 \in \rho(H(p,z))$ for all $z \in G_0$ with the possible exception of a finite number of points.

Proof :

(i) Let $z \in G$. By Lemma 10.5, there is a complex number $\zeta \neq 0$ belonging to the resolvent set of $H(p,z)$, and $[H(p,z) - \zeta]^{-1} \in \mathcal{B}_\infty$. By Lemma 9.11 and the Remark following the proof of Lemma 7.5, this implies that the spectrum of $H(p,z)$ consists only of eigenvalues of finite multiplicity, and that these eigenvalues do not accumulate at any finite point ξ in the complex plane. In particular we have either $0 \in \rho(H(p,z))$ or $0 \in \sigma_p(H(p,z))$. Also, by Lemma 10.6, there is at least one point $z_0 \in G$ such that $0 \in \rho(H(p,z_0))$.

(ii) Now let G_0 be a compact subset of G. Without loss of generality we may assume that $z_0 \in G_0$, where z_0 is a point such that $0 \in \rho(H(p,z_0))$, and that G_0 is connected.

Consider a fixed point $z_1 \in G_0$, and let $\zeta \neq 0$ be in $\rho(H(p,z_1))$. Also introduce the notation $R_\xi(z) = [H(p,z) - \xi]^{-1}$. One then has the identities

$$R_\zeta(z) = R_\zeta(z_1) \, [I + (H(p,z) - H(p,z_1)) \, R_\zeta(z_1)]^{-1} \tag{10.26}$$

and

$$R_\zeta(z) - R_\zeta(z_1) = -R_\zeta(z) \, [H(p,z) - H(p,z_1)] \, R_\zeta(z_1). \tag{10.27}$$

From Lemma 10.5 (i), we know that

$$(z - z_1)^{-1} [H(p,z) - H(p,z_1)] R_\zeta(z_1)$$

converges to a bounded operator as $z \to z_1$. Hence, given $\varepsilon > 0$, there is a number $\delta > 0$ such that $\|[H(p,z) - H(p,z_1)] R_\zeta(z_1)\| < \varepsilon$ for all $z \in G$ satisfying $|z-z_1| < \delta$. It then follows from Proposition 2.3 that the inverse on the r.h.s. of (10.26) exists and is bounded, so that $\zeta \in \rho(H(p,z))$ for all $z \in G$ satisfying $|z-z_1| < \delta$.

This together with (10.27) now implies that $z \mapsto R_\zeta(z)$ defines a holomorphic family of compact operators in the domain $\Delta_1(\delta) = \{z \in G \mid |z-z_1|<\delta\}$. Furthermore 0 is an eigenvalue of $H(p,z)$ if and only if $-\zeta^{-1}$ is an eigenvalue of $R_\zeta(z)$. Now by the analytic Fredholm theorem (the proof of which is contained in that of Lemma 7.5), one has the following alternative : either $-\zeta^{-1}$ is an eigenvalue of $R_\zeta(z)$ for all $z \in \Delta_1(\delta)$ or each compact subset of $\Delta_1(\delta)$ contains at most a finite number of points z for which $-\zeta^{-1}$ is an eigenvalue of $R_\zeta(z)$. In other words : Each point $z_1 \in G_0$ has a neighbourhood $N(z_1)$ such that either 0 is an eigenvalue of $H(p,z)$ for all $z \in N(z_1)$ or there are at most a finite number of points z in $N(z_1)$ such that $0 \in \sigma_p(H(p,z))$. Since G_0 is connected and compact we have : Either 0 is an eigenvalue of $H(p,z)$ for all $z \in G_0$, or there are only a finite number of points z in G_0 for which $0 \in \sigma_p(H(p,z))$. Since 0 is not an eigenvalue of $H(p,z)$ for some z_0 in G_0, the first alternative is excluded, and the Lemma is proven. ∎

<u>LEMMA 10.8</u> : Assume the hypotheses of Lemma 10.6 and set $H(k) = H_0(k) + Y$. Then the set of points $\{k \in \Gamma \mid 0 \notin \rho(H(k))\}$ is a set of Lebesgue measure zero.

<u>Proof</u> : It follows from Lemma 10.7 that, for each fixed $p \in \Gamma'$, the set of points $\{\kappa \in [0,1) \mid 0 \notin \rho(H(p,\kappa e_n))\}$ is a set of (linear) Lebesgue measure zero. This implies the assertion of the present lemma. ∎

<u>PROPOSITION 10.9</u> : Let $W : \mathbb{R}^n \to \mathbb{R}$ be periodic and such that $\|W\|_{s,C} < \infty$ for some s satisfying $s \geq 2$, $s > n-2$. Then the operator $H = K_0 + W$ has purely continuous spectrum.

Proof : Assume f is an eigenvector of H, i.e. $Hf = \lambda f$ for some $\lambda \in \mathbb{R}$. The potential $W'(x) = W(x) - \lambda$ also satisfies the assumptions of the Proposition, and we have $H'f = 0$, where $H' = K_0 + W'$. It follows that $(\mathcal{U}H'f)(k) = 0$ for almost all k. On the other hand, by (10.16),

$$(\mathcal{U}H'f)(k) = [H_0(k) + Y'] (\mathcal{U}f)(k) \equiv H'(k)(\mathcal{U}f)(k).$$

Hence

$$H'(k)(\mathcal{U}f)(k) = 0 \quad \text{a.e.} \quad \text{in } \Gamma.$$

Thus, for a. a. k, we either have $(\mathcal{U}f)(k) = 0$ or $0 \in \sigma(H'(k))$. Since by Lemma 10.8 the second alternative is possible only on a set of measure zero, we have $(\mathcal{U}f)(k) = 0$ for almost all k in Γ. Hence f is the zero vector, i.e. H cannot have any eigenvalues. ∎

In proposition 10.9, it is possible to prove a stronger result, namely that H has purely absolutely continuous spectrum. We shall indicate a proof of this in an Appendix to this chapter. For the moment we return to Schrödinger operators with (non-periodic) potential tending to zero at infinity and show how Proposition 10.9 can be used to prove absence of positive eigenvalues of such operators.

PROPOSITION 10.10 : Let $n \geq 2$, $V : \mathbb{R}^n \to \mathbb{R}$ such that

(α) $V \in L^s_{loc}(\mathbb{R}^n)$ for some s satisfying $s \geq 2$, $s > n-2$,

(β) $\lim |x|\, V(x) = 0$ as $|x| \to \infty$, uniformly in the direction $\omega = x|x|^{-1}$ of x. Define $H = K_0 + V$. Then H has no positive eigenvalues : $\sigma_p(H) \cap (0,\infty) = \emptyset$.

Proof :

(i) Assume that $f \in D(K_0) = D(H)$ is such that $Hf = \lambda f$ with $\lambda > 0$. We must show that $f = 0$. This is done in two steps : one first shows that, as a consequence of the hypothesis (β), f must have compact support. This will be done in (ii) - (vii) below following (Kato, [25]). Next, one may choose a periodic $W : \mathbb{R}^n \to \mathbb{R}$ such that $W(x) = V(x)$ for all x in the support of f. If $H' = K_0 + W$, then $H'f = Hf = \lambda f$, since $Vf = Wf$. Hence f is an eigenvector of the periodic Schrödinger operator H'. By the hypothesis (α) and Proposition 10.9, we conclude that $f = 0$.

(ii) To prove that f has compact support, we shall use the following two facts :

(a) There is a sequence $\{f_k\}$ in $C_0^\infty(\mathbb{R}^n)$ such that $\|f - f_k\| \to 0$ and $\|K_0 f - K_0 f_k\| \to 0$ as $k \to \infty$ (see [K], p. 300).

(b) If $\mathcal{H}_0$ is a Hilbert space, g and g_k in $L^2(\mathbb{R} ; \mathcal{H}_0, dx)$ and $\|g-g_k\| \to 0$ as $k \to \infty$, there is a subsequence g_{k_i} such that, for almost all $x \in \mathbb{R}$, $\|g_{k_i}(x) - g(x)\|_{\mathcal{H}_0} \to 0$ as $i \to \infty$ (see [DS, vol. I]).

Notice that the sequence $\{f_k\}$ in (a) also has the property that $\|Hf_k - Hf\| = \|Hf_k - \lambda f\| \to 0$ and $\|P_i f_k - P_i f\| \to 0$ as $k \to \infty$, since for example $\|V(f_k - f)\| \le \|V(K_0 + 1)^{-1}\| \, \|(K_0 + 1)(f_k - f)\| \to 0$.

(iii) We introduce spherical polar coordinates (r,ω) in $\mathbb{R}^n$, where $r = |x|$ is the length of the vector x and $\omega = |x|^{-1}x \in S^{n-1}$ a unit vector along x. To each $g \in L^2(\mathbb{R}^n)$ we associate the following family $\{\theta_g(r)\}$ of vectors in $L^2(S^{n-1})$, defined for almost all $r \in (0,\infty)$:

$$\theta_g(r)(\omega) = r^{\frac{n-1}{2}} g(r\omega) . \tag{10.28}$$

We then have

$$\|g\|^2_{L^2(\mathbb{R}^n)} = \int_0^\infty dr \, \|\theta_g(r)\|^2_{L^2(S^{n-1})} . \tag{10.29}$$

In what follows, the subscripts on norms and scalar products will be omitted.

Using the expression for the Laplacian in spherical polar coordinates, one finds that, for $g \in C_0^\infty(\mathbb{R}^n)$,

$$\theta_{Hg}(r) = - \frac{d^2}{dr^2} \theta_g(r) - r^{-2} \Delta_S \theta_g(r) + \frac{1}{4} (n-1)(n-3) r^{-2} \theta_g(r)$$

$$+ V(r.) \theta_g(r) , \tag{10.30}$$

where Δ_S is the spherical Laplacian acting in $L^2(S^{n-1})$. We shall not need the explicit form of Δ_S but use the fact that it is an (unbounded) *negative* self-adjoint operator. The vector $\frac{d^2}{dr^2}\theta_g(r)$ may be defined as a strong derivative in $L^2(S^{n-1})$ or simply in terms of the radial derivatives of g(x) by using (10.28).

We also define the following vectors in $L^2(S^{n-1})$: If $\{f_k\}$ is the sequence defined in (ii,a) we set

$$\theta_k(r) = \theta_{f_k}(r) = r^{\frac{n-1}{2}} f_k(r.) \tag{10.31}$$

and, for each integer $\nu \geq 0$:

$$\theta_{k,\nu}(r) = r^\nu \theta_k(r) = r^{\nu+\frac{n-1}{2}} f_k(r.) . \tag{10.32}$$

Using (10.30), we find that $\theta_{k,\nu}$ satisfies the following differential equation :

$$\theta''_{k,\nu} = 2\nu r^{-1} \theta'_{k,\nu} - r^\nu \theta_{Hf_k} + V(r.)\,\theta_{k,\nu}$$

$$- r^{-2} [\nu(\nu+1) - \tfrac{1}{4}(n-1)(n-3) + \Delta_S]\,\theta_{k,\nu} . \tag{10.33}$$

Since the eigenvector f of H is in $D(K_o)$, i.e. $f = (K_o + 1)^{-1} f_o$ for some $f_o \in L^2(\mathbb{R}^n)$, we obtain from Proposition 8.2 and (8.4) that $r \mapsto \theta_f(r)$ is strongly continuous from $(0,\infty)$ to $L^2(S^{n-1})$ and that

$$\|\theta_f(r)\| \leq M < \infty \quad \text{for all} \quad r \in (0,\infty). \tag{10.34}$$

(iv) Let $\Sigma(r)$ be the multiplication operator in $L^2(S^{n-1})$ by the function $\sigma_r(\omega) = r\,V(r\omega) + \frac{1}{4}(n-1)(n-3)\,r^{-1}$. Then by hypothesis (β) we may choose a number R $\in (0,\infty)$ such that

$$\|\Sigma(r)\| < \lambda^{\frac{1}{2}} \quad \text{for all} \quad r \geq R \ . \tag{10.35}$$

We then have for each $r \geq R$:

$$\|\theta_k'(r) + \Sigma(r)\,\theta_k(r)\|^2 \leq \|\theta_k'(r)\|^2 + \lambda\|\theta_k(r)\|^2$$

$$+ (\theta_k',\Sigma(r)\theta_k) + (\Sigma(r)\theta_k,\theta_k') \ . \tag{10.36}$$

Finally we define functions ϕ_k and $\psi_{k,\nu}$ from $(0,\infty)$ to $\mathbb{R}$ as follows :

$$\phi_k(r) = \|\theta_k'(r)\|^2 + \lambda\,\|\theta_k(r)\|^2 + r^{-2}(\theta_k(r),\Delta_S\,\theta_k(r)) \ , \tag{10.37}$$

$$\psi_{k,\nu}(r) = \|\theta_{k,\nu}'(r)\|^2 + [\lambda - \lambda R r^{-1} + \nu(\nu+1)\,r^{-2}]\,\|\theta_{k,\nu}(r)\|^2$$

$$+ r^{-2}\,(\theta_{k,\nu}(r)\ ,\ \Delta_S\,\theta_{k,\nu}(r))$$

$$= r^{2\nu}\,\{\|\theta_k'(r) + \nu\,r^{-1}\,\theta_k(r)\|^2$$

$$+ [\lambda - \lambda R r^{-1} + \nu(\nu+1)\,r^{-2}]\,\|\theta_k(r)\|^2 + r^{-2}(\theta_k,\Delta_S\,\theta_k)\}. \tag{10.38}$$

Using (10.30), (10.33), the inequality (10.36) and the positivity of $-\Delta_S$, one then finds that, for $r \geq 2R$:

$$\frac{d}{dr}[r\,\phi_k(r)] = \|\theta_k'\|^2 + \lambda\,\|\theta_k\|^2 + \lambda\, r(\theta_k', \theta_k) + \lambda\, r(\theta_k, \theta_k')$$

$$- r(\theta_k', \theta_{Hf_k}) - r(\theta_{Hf_k}, \theta_k')$$

$$+ (\theta_k', \Sigma(r)\,\theta_k) + (\Sigma(r)\,\theta_k, \theta_k') - r^{-2}(\theta_k, \Delta_S\,\theta_k)$$

$$\geq r(\theta_k', \lambda\theta_k - \theta_{Hf_k}) + r(\lambda\theta_k - \theta_{Hf_k}, \theta_k') \qquad (10.39)$$

and

$$\frac{d}{dr}[r^2\,\psi_{k,\nu}(r)] = 2r(1+2\nu)\,\|\theta_{k,\nu}'\|^2 + \lambda(2r-R)\,\|\theta_{k,\nu}\|^2$$

$$+ r(\theta_{k,\nu}', \Sigma(r)\,\theta_{k,\nu}) + r(\Sigma(r)\,\theta_{k,\nu}, \theta_{k,\nu}')$$

$$+ r^2(\theta_{k,\nu}', \lambda\theta_{k,\nu} - r^{\nu}\theta_{Hf_k}) + r^{-2}(\lambda\theta_{k,\nu} - r^{\nu}\theta_{Hf_k}, \theta_{k,\nu}')$$

$$- \lambda\, Rr(\theta_{k,\nu}', \theta_{k,\nu}) - \lambda\, Rr(\theta_{k,\nu}, \theta_{k,\nu}')$$

$$\geq r(1+4\nu)\,\|\theta_{k,\nu}'\|^2 + \lambda(r-R)\,\|\theta_{k,\nu}\|^2$$

$$- 2r^2\,\|\theta_{k,\nu}'\|\,\|\lambda\theta_{k,\nu} - r^{\nu}\theta_{Hf_k}\| - 2\lambda Rr\,\|\theta_{k,\nu}\|\,\|\theta_{k,\nu}'\|$$

$$\geq r\,\{(1+4\nu)\,\|\theta_{k,\nu}'\|^2 - 2\lambda R\,\|\theta_{k,\nu}\|\,\|\theta_{k,\nu}'\| + \tfrac{1}{2}\,\lambda\|\theta_{k,\nu}\|^2\}$$

$$- 2r^2\,\|\theta_{k,\nu}'\|\,\|\lambda\theta_{k,\nu} - r^{\nu}\theta_{Hf_k}\|. \qquad (10.40)$$

The first term on the r.h.s. is positive for all $r \geq 2R$ and all $\nu \geq \nu_0 \equiv \frac{1}{2}\,\lambda\, R^2$. Hence

$$\frac{d}{dr}\,[r^2\psi_{k,\nu}(r)] \geq -\,2r^2\,\|\theta'_{k,\nu}\|\,\|\lambda\theta_{k,\nu} - r^{\nu}\theta_{Hf_k}\| \tag{10.41}$$

provided that $r \geq 2R$ and $\nu \geq \nu_o$.

(v) Let $g \in C_o^\infty(\mathbb{R}^n)$. We get upon using (10.30) for V = 0 and integrating by parts that

$$\sum_{j=1}^{n} \|P_j\ g\|^2 = (g,K_o g)$$
$$= \int_0^\infty dr\ \{\|\theta'_g\|^2 - r^{-2}\ (\theta_g,\Delta_S\theta_g) + \frac{1}{4}\,(n-1)(n-3)\ r^{-2}\ \|\theta_g\|^2\ . \tag{10.42}$$

Taking $g = f_k - f_\ell$, we have $(f_k - f_\ell\ ,\ K_o(f_k - f_\ell)) \to 0$ as $k,\ell \to \infty$. Since each term in the integrand in (10.42) is non-negative, this means that each of the three families $w_k = \{\theta'_k(r)\}, \{r^{-1}(-\Delta_S)^{1/2}\ \theta_k(r)\}$ and $\{r^{-1}\theta_k(r)\}$ defines a Cauchy sequence in $L^2((0,\infty)\ ;\ L^2(S^{n-1}))$. We denote the limit vector of $\{\theta'_k(r)\}$ by $h = \{\theta_h(r)\}$ and that of $\{r^{-1}(-\Delta_S)^{1/2}\ \theta_k(r)\}$ by $e = \{e(r)\}$. We also have that $f_k = \{\theta_k(r)\}$ converges in $L^2((0,\infty)\ ;\ L^2(S^{n-1}))$ to $f = \{\theta_f(r)\}$ and that $Vf_k = \{V(r.)\ \theta_k(r)\}$ converges to $Vf = \{V(r.)\ \theta_f(r)\}$.

By (ii.b) there is a sequence $\{k_i\}$ of integers ($k_i \to \infty$) and a null set Δ_o in $(2R,\infty)$ such that one has for each $r \in (2R,\infty)\backslash\Delta_o$, in the sense of strong convergence in $L^2(S^{n-1})$:

$$\theta'_{k_i}(r) \to \theta_h(r)\quad ,\quad \theta_{k_i}(r) \to \theta_f(r)$$

$$V(r.)\ \theta_{k_i}(r) \to V(r.)\ \theta_f(r)$$

and

$$(-\Delta_S)^{\frac{1}{2}}\ \theta_{k_i}(r) \to r\ e(r)\ .$$

It follows that, for $r \in (2R,\infty)\setminus\Delta_0$:

$$\lim_{i\to\infty} \phi_{k_i}(r) \equiv \phi(r) = \|\theta_h(r)\|^2 + \lambda \|\theta_f(r)\|^2 - \|e(r)\|^2 \tag{10.43}$$

and

$$\lim_{i\to\infty} \psi_{k_i,\nu}(r) \equiv \psi_\nu(r) = r^{2\nu} \{\|\theta_h(r) + \nu r^{-1}\theta_f(r)\|^2$$
$$+ [\lambda - \lambda Rr^{-1} + \nu(\nu+1) r^{-2}] \|\theta_f(r)\|^2 - \|e(r)\|^2\}. \tag{10.44}$$

We obtain in particular that, for each $s \geq 2R$, $\lim_{i\to\infty} \int_s^\infty \phi_{k_i}(r)\, dr = \int_s^\infty \phi(r)\, dr$. Since $\int_s^\infty \phi_k(r)\, dr \leq c_n < \infty$ for all k, we have

$$\int_s^\infty \phi(r)\, dr < \infty. \tag{10.45}$$

From now on we label the preceding subsequence $\{k_i\}$ of $\mathbb{Z}^+$ simply by the letter i.

(vi) Let $r,s \in (2R,\infty)\setminus\Delta_0$ and $r > s$. Then (10.39) and (10.42) imply that

$$r\, \phi_i(r) - s\, \phi_i(r) \geq -2 \int_s^r u\, \|\theta_i'(u)\|\, \|\theta_{Hf_i}(u) - \lambda\theta_i(u)\|\, du$$

$$\geq -2r(f_i,K_0 f_i)\, \|Hf_i - \lambda f_i\|.$$

Since the r.h.s. converges to zero as $i \to \infty$, we obtain

$$r\, \phi(r) \geq s\, \phi(s) \quad \text{if} \quad r > s \geq 2R\ ,\ r,s \notin \Delta_0. \tag{10.46}$$

Similarly we obtain from (10.40) that

$$r^2\, \psi_\nu(r) \geq s^2\, \psi_\nu(s) \quad \text{if} \quad r > s \geq 2R\ ,\ r,s \notin \Delta_0\ ,\ \nu \geq \nu_0. \tag{10.47}$$

Notice that (10.46) implies that

$$\int_s^\infty \phi(r)\, dr \geq s\, \phi(s) \int_s^\infty r^{-1}\, dr \, .$$

Since the integral on the l.h.s. is less than $+\infty$ by (10.45), we must have

$$\phi(s) \leq 0 \quad \text{for each} \quad s \in (2R,\infty)\backslash\Delta_0 \, . \tag{10.48}$$

(vii) We now show that the support of the eigenvector f of H is contained in some ball $\{x \mid |x| \leq M\}$. Assume the contrary. Then there is a $t \in (2R,\infty)$ such that $\theta_f(t) \neq 0$. Since $r \mapsto \|\theta_f(r)\|$ is continuous, we may assume that $t \notin \Delta_0$. It then follows from the expression (10.44) for $\psi_\nu(t)$ that there is a ν_1 such that $\psi_\nu(t) > 0$ for all $\nu \geq \nu_1$. By combining this with (10.47), we find $\psi_\nu(r) > 0$ for each $\nu \geq \kappa \equiv \max(\nu_0,\nu_1)$ and each $r \in [t,\infty)\backslash\Delta_0$.

Now choose $M \geq t$ such that $\kappa(2\kappa+1)r^{-2} - \lambda R r^{-1} < 0$ for $r \geq M$. If the support of f is not contained in the ball $\{x \mid |x| \leq M\}$, then $0 < \int_{2M}^\infty \|\theta_f(s)\|^2 ds \leq \|f\|^2 < \infty$, so that there are numbers $\rho, r \notin \Delta_0$ such that $M \leq \rho < r < \infty$ and $\|\theta_f(\rho)\| > \|\theta_f(r)\| \geq 0$. Hence

$$\begin{aligned} 0 > \lim_{i\to\infty} (\|\theta_i(r)\|^2 - \|\theta_i(\rho)\|^2) &= \lim_{i\to\infty} \int_\rho^r \frac{d}{du} \|\theta_i(u)\|^2 \, du \\ &= \lim_{i\to\infty} \int_\rho^r \{(\theta_i'(u) \, , \, \theta_i(u)) + (\theta_i(u) \, , \, \theta_i'(u))\} \, du \\ &= \lim_{i\to\infty} \{(F_{\rho r}\, w_i \, , \, F_{\rho r}\, f_i) + (F_{\rho r}\, f_i \, , \, F_{\rho r}\, w_i)\} \\ &= \int_\rho^r \{(\theta_h(u) \, , \, \theta_f(u)) + (\theta_f(u) \, , \, \theta_h(u))\} \, du \, , \end{aligned} \tag{10.49}$$

where $F_{\rho r}$ is the following operator :

$$(F_{\rho r} g)(x) = \chi_{(\rho,r)}(|x|)\ g(x)\ ,$$

with $\chi_{(\rho,r)}(s) = 1$ if $s \in (\rho,r)$ and $\chi_{(\rho,r)}(s) = 0$ otherwise. (10.49) implies the existence of a number $s \in (\rho,r)$ such that $s \notin \Delta_0$ and $(\theta_h(s)\ ,\ \theta_f(s)) + (\theta_f(s)\ ,\ \theta_h(s)) < 0$. Hence

$$\|\theta_h(s) + \nu s^{-1}\ \theta_f(s)\|^2 \leq \|\theta_h(s)\|^2 + \nu^2\ s^{-2}\ \|\theta_f(s)\|^2\ .$$

This in turn implies together with (10.43) and (10.44) that

$$\psi_\kappa(s) \leq s^{2\kappa}\ \phi(s)\ .$$

We have seen above that $\psi_\kappa(s) > 0$; hence $\phi(s) > 0$, which contradicts (10.48). Therefore there is no $\rho > M$ such that $\theta_f(\rho) \neq 0$, which proves our claim that supp $f \subseteq \{x \mid |x| \leq M\}$. ■

APPENDIX : ABSOLUTE CONTINUITY OF PERIODIC SCHRÖDINGER OPERATORS.

The purpose of this appendix is to improve the result of Proposition 10.9 :

PROPOSITION 10.11 : Let W be periodic and such that $\|W\|_{s,C} < \infty$ for some s satisfying $s \geq 2$, $s > n-2$. Then the operator $H = K_o + W$ has purely absolutely continuous spectrum.

For the proof, we write $G = L^2(\Gamma ; \ell^2)$ as follows :

$$G = L^2(\Gamma ; \ell^2) = L^2(\Gamma' ; L^2([0,\ell] ; \ell^2)) . \tag{10.50}$$

In other words, each $f \in L^2(\Gamma ; \ell^2)$ is viewed as a square integrable function from Γ' to $L^2([0,\ell] ; \ell^2)$. We use the letter p to denote points in Γ' and the letter t to denote points in $\mathbb{R}$. We denote by $H(p,t)$ the operator

$$H(p,t) = H_o(p,te_n) + Y$$

acting in ℓ^2, and by $K(p)$ the operator

$$[K(p)g](t) = H(p,te_n)\ g(t)$$

acting in $L^2([0,\ell] ; \ell^2)$ (i.e. $g(t) \in \ell^2$ for each $t \in [0,\ell]$).

The idea of the proof of Proposition 10.11 is to show first that each $K(p)$ has purely absolutely continuous spectrum as an operator in $L^2([0,\ell] ; \ell^2)$, which will be done in Lemmas 10.12 through 10.14. From this one can then deduce that H has purely absolutely continuous spectrum by noticing that

$$(\mathcal{U}Hf)(p) = K(p)(\mathcal{U}f)(p) \qquad (p \in \Gamma') , \tag{10.51}$$

where $(\mathcal{U}f)(p) \in L^2([0,\ell] ; \ell^2)$ is defined in a obvious way. This second step of the proof will given in Lemma 10.15.

In three of the four lemmas below we shall not refer explicitly to the situation just described but rather stay on an abstract level. It is easily verified that the situation of Proposition 10.11 is covered by these abstract results.

LEMMA 10.12 : Let $A(t)$ be an analytic family of self-adjoint operators with compact resolvent, for $t \in I \subset \mathbb{R}$. Suppose that, at $t_0 \in I$, $\{\lambda_j(t_0)\}$ are all the eigenvalues of $A(t_0)$, counting multiplicities. In particular, suppose $\mu_0 = \lambda_j(t_0) = \ldots = \lambda_{j+k-1}(t_0)$ occurs with multiplicity k. Then, there are an interval U about μ_0 and an interval $J \subset I$ about t_0, with the following properties : there exist real analytic function $\lambda_j(t),\ldots,\lambda_{j+k-1}(t)$, $t \in J$, such that, for $t \in J$, all the eigenvalues of $A(t)$ in U are $\{\lambda_j(t),\ldots,\lambda_{j+k-1}(t)\}$. (We remark not all the $\lambda_\nu(t)$, $j \le \nu \le j+k-1$ need be distinct).

Proof : To prove this lemma, we note first that it is elementary in case the Hilbert space on which $A(t)$ acts is of (finite) dimension k. In fact, solving the characteristic equation, one sees that the k eigenvalues of $A(t)$ in this case can be written in the form

$$\lambda_j(t) = \mu_0 + \sum_{\nu=1}^{k} \alpha_{j\nu}(t-t_0)^{\frac{\nu}{\mu}} \tag{10.52}$$

for some integer μ. However, the requirement that all eigenvalues $\lambda_j(t)$ be *real* implies that all the exponents in (10.52) are *integers*, which implies that the functions λ_j are analytic. Thus, it remains to reduce the general case to this special case.

First, if γ is a sufficiently small circle in the complex plane $\mathbb{C}$, centered at μ_0, then, for t close to t_0

$$P(t) = (2\pi i)^{-1} \int_\gamma (A(t) - z)^{-1} dz$$

is the orthogonal projection onto the k-dimensional space spanned the eigenspaces associated with all the eigenvalues of A(t) which lie in U, a small neighbourhood of μ_0, encircled by γ. It will suffice to produce $\lambda_j(t),\ldots,\lambda_{j+k-1}(t)$, real analytic, which are the eigenvalues of B(t), the restriction of A(t) to the k-dimensional space

$$N(t) = \text{range } P(t) .$$

Now the operators $U(t) = P(t_0)\, P(t) + [I - P(t_0)]\, [I - P(t)]$ form an analytic family of invertible operators, and U(t) maps $N(t)$ onto $N(t_0)$. Hence

$$\tilde{B}(t) = U(t)\, B(t)\, U(t)^{-1}$$

is an analytic family of self-adjoint operators on the fixed k-dimensional space $N(t_0)$, so we know it has k analytic families of eigenvalues $\lambda_j(t),\ldots,\lambda_{j+k-1}(t)$, which are necessarily the eigenvalues of B(t).∎

<u>LEMMA 10.13</u> : Let W be as in Proposition 10.11, $p \in \Gamma'$ and $A(t) = H(p,t)$. Then the family $\{A(t)\}_{t\in\mathbb{R}}$ has the following property : there are real analytic functions $\lambda_j(t)$ such that, for each t, *all* the eigenvalues of A(t), with multiplicities taken into account, are $\{\lambda_j(t)\}$. Furthermore none of the functions λ_j is constant.

<u>Proof</u> : For any $t_0 \in \mathbb{R}$ and any N, Lemma 10.12 shows that there is an interval J about t_0 in $\mathbb{R}$ such that $\{\lambda_j(t_0) : 1 \le j \le N\}$ have analytic continuations to eigenvalues of A(t), provided $\lambda_j(t_0)$ are eigenvalues of $A(t_0)$. Furthermore, if $\{\lambda_j(t_0) : 1 \le j \le N\}$ is all the eigenvalues of $A(t_0)$ on [-R,R], then for a small enough interval J, $\{\lambda_j(t) : 1 \le j \le N\}$ will contain all the eigenvalues of A(t) in the interval [-R-1,R+1], for each $t \in J$.

Consequently, it remains to show that each $\lambda_j(t)$ can be continued as an analytic function for all real t, and for this it suffices to show that no $\lambda_j(t)$ blows up for finite t. In fact, from

$$(\frac{d}{dt} H(p,t)g)_m = 2\ell^2(t + m_n)g_m , \tag{10.53}$$

which follows from (10.12), we claim that, on any interval $I \supset J$ on which $\lambda_j(t)$ is analytic

$$|\frac{d}{dt} \lambda_j(t)| \le c_j[|\lambda_j(t)| + 1 + |t|] , \tag{10.54}$$

which certainly implies no blow-up for finite t.

To show (10.54), we can content ourselves with proving it on a dense subset of I. It is clear that, except for a discrete subset S of I, we can find an analytic family $e_j(t)$ with $\|e_j(t)\| = 1$, $A(t)\, e_j(t) = \lambda_j(t)\, e_j(t)$. In fact, with more work, along the lines of the proof of Lemma 10.12, one can produce $e_j(t)$ analytic on all of I, but we do not need this here. Now, since $\|e_j\| = 1$ implies $(e_j', e_j) + (e_j, e_j') \equiv 0$, we have, on $I \backslash S$ $(e_j'(t) \equiv \frac{d}{dt}\, e_j(t))$:

$$|\frac{d\lambda_j}{dt}| = |(A'e_j,e_j) + (Ae_j',e_j) + (Ae_j,e_j')|$$

$$= |(A'e_j,e_j) + \lambda_j[(e_j',e_j) + (e_j,e_j')]|$$

$$= |(A'e_j,e_j)| \le \|A'e_j\| . \tag{10.55}$$

If P_n denotes the multiplication operator by m_n, then $\|P_n[A(t) - i]^{-1}\| \le c < \infty$ for all $t \in I$. (see [K], ch. VII 2.1). Thus (10.55) and (10.53) imply that (10.54) holds on $I\backslash S$, and hence by continuity, on all of I. The non-constancy of λ_j follows from Lemma 10.7 by using an argument similar to that of the proof of Proposition 10.9. ∎

LEMMA 10.14 : Let $\mathcal{H}_0$ be an infinite dimensional Hilbert space and $\mathcal{H} = L^2([a,b],\mathcal{H}_0 \,;\, dx)$. Let A be an operator of the form

$$(Af)(x) = A(x)\, f(x) \qquad f \in \mathcal{H},\ x \in [a,b],$$

where each $A(x)$ is a self-adjoint operator in $\mathcal{H}_0$. Suppose there are $\mathcal{H}_0$-valued measurable functions $\{e_n(x)\}_{i=1}^{\infty}$ forming an orthonormal basis of $\mathcal{H}_0$ for each x such that

$$A(x)\, e_n(x) = \lambda_n(x)\, e_n(x) \quad ,$$

with each λ_n real-valued, analytic on (a,b) and continuous on [a,b]. Assume no λ_n is constant. Then A has purely absolutely continuous spectrum.

Proof : Letting the $\{e_n(x)\}$ define an isomorphism of $\mathcal{H}_0$ and ℓ^2 for each x, we see that A is unitarily equivalent to an operator A_0 on $L^2([a,b] \,;\, \ell^2)$ given by

$$[(A_0 f)(x)]_n = \lambda_n(x)\ [f(x)]_n \quad , \tag{10.56}$$

where $\{[f(x)]_n\} \in L^2([a,b] \,;\, \ell^2)$. It remains to show that A_0 has purely absolutely continuous spectrum, given the analyticity and non-constancy hypotheses on λ_n. Now, letting $A_{o,n}$ be the multiplication operator by $\lambda_n(x)$ in $L^2(a,b)$, one sees that A_0 is the direct sum of the operators $A_{o,n}$: $A_0 = \oplus_{n=1}^{\infty} A_{o,n}$. Hence it suffices to show that each $A_{o,n}$ has purely absolutely continuous spectrum. But this is easily seen, upon partitioning (a,b) into at most denumerably many intervals on the interior of each of which $\lambda_n(x)$ is strictly monotonic, with non-vanishing derivative. ∎

LEMMA 10.15 : Let $\{M,\mu\}$ be a separable σ-finite measure space, $\mathcal{H}_0$ an infinite-dimensional Hilbert space and $\mathcal{H} = L^2(M ; \mathcal{H}_0, d\mu)$. For each $x \in M$, let $B(x)$ be a self-adjoint operator in $\mathcal{H}_0$, and assume that $(f,B(.)g)$ is μ-measurable for all $f,g \in \mathcal{H}_0$. Let B be the following self-adjoint operator in $\mathcal{H}$:

$$(Bf)(x) = B(x)\, f(x) \tag{10.57}$$

with

$$D(B) = \{f \in \mathcal{H} \mid f(x) \in D(B(x)) \quad \text{a.e. and}$$

$$\int_M \|B(x)\, f(x)\|^2_{\mathcal{H}_0}\, d\mu(x) < \infty\}. \tag{10.58}$$

Assume that each $B(x)$ has purely absolutely continuous spectrum. Then B has purely absolutely continuous spectrum.

Proof : If $\phi : \mathbb{R} \to \mathbb{C}$ is bounded and measurable, one has (see Problem 10.2)

$$[\phi(B)\, f]\,(x) = \phi(B(x))\, f(x). \tag{10.59}$$

In particular, if $\{E_\lambda\}$ denotes the spectral family of B and $\{E^x_\lambda\}$ that of $B(x)$:

$$(f,E_\lambda f) = \int_M (f(x),\, E^x_\lambda\, f(x))_{\mathcal{H}_0}\, d\mu(x)\ . \tag{10.60}$$

By the definition of absolute continuity in Chapter 4, there is, for each fixed x, a non-negative function $\psi_{f,x} \in L^1(\mathbb{R})$ such that

$$(f(x)\ ,\ E_\lambda^x\ f(x))_{\mathcal{H}_0} = \int_{-\infty}^{\lambda} \psi_{f,x}(s)\ ds\ .$$

Inserting this into (10.60) and interchanging the order of integration by using Fubini's theorem, we see that

$$(f\ ,\ E_\lambda f) = \int_{-\infty}^{\lambda} ds\ \theta_f(s)\ ,$$

where $\theta_f(s) = \int_M \psi_{f,x}(s)\ d\mu(x) \in L^1(\mathbb{R})$. Consequently $m_f(\Delta) = (f\ ,E_\Delta\ f)$ defines an absolutely continuous measure for each $f \in \mathcal{H}$, showing that $\mathcal{H} = \mathcal{H}_{ac}(B)$. ∎

11 Eigenfunction expansions

We begin with two examples.

(1) Let A be a self-adjoint operator with pure point spectrum. Then there exists an orthonormal basis $\{e_k\}$ of $\mathcal{H}$ formed of eigenvectors of A, i.e.

$$A\, e_k = \lambda_k\, e_k \quad , \quad \text{for some } \lambda_k \in \mathbb{R} ,$$

$$(e_j\, , e_k) = \delta_{jk} \quad ,$$

$$\{e_k\} \quad \text{span } \mathcal{H} .$$

Let f be an arbitrary vector in $\mathcal{H}$. Then f can be expanded in the chosen basis $\{e_k\}$ of eigenvectors of A :

$$f = \sum_k \alpha_k\, e_k \quad , \tag{11.1}$$

with

$$\alpha_k = (e_k\, , f) . \tag{11.2}$$

More precisely $f = \operatorname*{s-lim}_{N\to\infty} \sum_{k=1}^{N} \alpha_k\, e_k$

and

$$\|f\|^2 = \sum_k |\alpha_k|^2 = \sum_k |(e_k\, , f)|^2 . \tag{11.3}$$

(2) The operator K_o in $L^2(\mathbb{R}^n)$ has absolutely continuous spectrum, hence no eigenvectors in $\mathcal{H} = L^2(\mathbb{R}^n)$. However one may view K_o as a differential operator, $K_o = -\Delta$, and the latter has eigenfunctions, namely the so-called *planes waves*

$$\psi_k^0(x) = (2\pi)^{-\frac{n}{2}} e^{ik.x} \quad .$$

These are C^∞-functions of the variable x that verify $(-\Delta\psi_k^0)(x) = k^2\, \psi_k^0(x)$ but which are not square-integrable over $\mathbb{R}^n$.

Every vector $f \in L^2(\mathbb{R}^n)$ can be expanded in these eigenfunctions, by the theory of the Fourier transformation :

$$f(x) = \int dk\, \psi_k^0(x)\, \tilde{f}(k) \quad , \tag{11.1'}$$

with

$$\tilde{f}(k) = \int dx\, \overline{\psi_k^0(x)}\, f(x) \quad , \tag{11.2'}$$

and

$$\|f\|^2 = \int dk\, |\tilde{f}(k)|^2 \quad . \tag{11.3'}$$

These equations are the exact analogue of (11.1) - (11.3). Some caution is required in defining these integrals. (11.1') should be written more carefully as

$$f(x) = \text{l.i.m} \int dk\, \psi_k^0(x)\, \tilde{f}(k) \quad ,$$

or equivalently

$$f = \underset{N\to\infty}{\text{s-lim}}\, f_N \quad , \quad f_N(x) = \int_{|x|\leq N} dk\, \psi_k^0(x)\, \tilde{f}(k) \quad ,$$

and similarly for (11.2').

The Fourier transformation also gives the spectral transformation of K_0, i.e. the unitary operator $\mathcal{U}_0 : L^2(\mathbb{R}^n) \to L^2([0,\infty), L^2(S^{(n-1)}))$:

$$(\mathcal{U}_0 f)_\lambda (\omega) = 2^{-\frac{1}{2}} \lambda^{\frac{n-2}{4}} \tilde{f}(\sqrt{\lambda}\omega) \qquad [\omega \in S^{(n-1)}]$$

is such that $(\mathcal{U}_0 K_0 f)_\lambda = \lambda(\mathcal{U}_0 f)_\lambda$ for $f \in D(K_0)$, see Chapter 4.

The aim of this chapter is to obtain similar relations for the perturbed operator $H = K_0 + V$, for a class of functions V. We already know that $\mathcal{H} = \mathcal{H}_p(H) \oplus \mathcal{H}_{ac}(H)$. The subspace $\mathcal{H}_p(H)$ is treated as in Example (1) and will not be discussed. For the absolutely continuous part, one looks for eigenfunctions $\psi_k(x)$ satisfying $[-\Delta + V(x)]\, \psi_k(x) = k^2\, \psi_k(x)$ and tries to obtain relations analogous to (11.1') - (11.3') for $f \in \mathcal{H}_{ac}(H)$ as well as spectral transformations of H. Since we have already studied the wave operators, it is in fact easy to get spectral transformations of H. Throughout this chapter we assume that $n > 3$.

PROPOSITION 11.1 : Let $H = K_0 + V$ in $L^2(\mathbb{R}^n)$, $n \geq 3$. Assume that

$$V(x) = A(x)\, B(x) = (1 + |x|)^{-\alpha}\, [B_1(x) + B_2(x)]$$

with $\alpha > 1$, $B_1(.) \in L^{p_1}(\mathbb{R}^n) \cap L^{p_2}(\mathbb{R}^n)$, $p_1 < n < p_2$ and $|B_2(x)| \leq c(1 + |x|)^{-\frac{1}{2}-\nu}$ $(\nu > 0)$. Define $\mathcal{U}_\pm$ by

$$\mathcal{U}_\pm = \mathcal{U}_0\, \Omega_\pm^* .$$

$\mathcal{U}_\pm$ are spectral transformations of H. More precisely : $\mathcal{U}_\pm$ are unitary maps from $\mathcal{H}_{ac}(H)$ onto $G_0 \equiv L^2([0,\infty), L^2(S^{(n-1)}))$ such that for $f \in D(H) \cap \mathcal{H}_{ac}(H)$:

$$(\mathcal{U}_\pm H f)_\lambda = \lambda(\mathcal{U}_\pm f)_\lambda .$$

Furthermore, if f is in the range of A and $\lambda \in (0,\infty)\backslash\Gamma_0$, then

$$(\mathcal{U}_\pm f)_\lambda = M_A(\lambda)(I + W_{\lambda\pm io})^{-1} A^{-1} f.$$

Remark : We wish to point out that

(i) The space G_0 giving the spectral representation of H_{ac} is the same as that giving the spectral representation of K_0.

(ii) our construction of the spectral transformation of H_{ac} involves the knowledge of that of K_0.

Proof :

(i)
$$\mathcal{U}_\pm^* \mathcal{U}_\pm = \Omega_\pm \mathcal{U}_0^* \mathcal{U}_0 \Omega_\pm^*$$

$$= \Omega_\pm \Omega_\pm^* = E_{ac}(H) ,$$

since the wave operators are complete.
Also

$$\mathcal{U}_\pm \mathcal{U}_\pm^* = \mathcal{U}_0 \Omega_\pm^* \Omega_\pm \mathcal{U}_0^* = \mathcal{U}_0 \mathcal{U}_0^* = I_{G_0} .$$

Thus $\mathcal{U}_\pm$ are unitary operators from $\mathcal{H}_{ac}(H)$ onto G_0.

Let $f \in D(H) \cap \mathcal{H}_{ac}(H)$. Using Proposition 6.6, we obtain

$$(\mathcal{U}_\pm H f)_\lambda = (\mathcal{U}_0 \Omega_\pm^* H f)_\lambda = (\mathcal{U}_0 K_0 \Omega_\pm^* f)_\lambda$$

$$= \lambda(\mathcal{U}_0 \Omega_\pm^* f)_\lambda = \lambda(\mathcal{U}_\pm f)_\lambda ,$$

which shows that $\mathcal{U}_\pm$ are spectral transformations of H.

(ii) Let $f \in A\mathcal{H}$, $g \in \mathcal{S}(\mathbb{R}^n)$, Δ a compact interval in $(0,\infty)\backslash\Gamma_0$. From the Lippman-Schwinger equation (Proposition 6.5) :

$$(f, \Omega_\pm E^0_\Delta g) = (f, E^0_\Delta g) - \lim_{\varepsilon\to+0} \int_\Delta (f, (H-\lambda\pm i\varepsilon)^{-1} V\, dE^0_\lambda g).$$

We now write $V = BA$ and use (7.22), viz.

$$I - B(H-z)^{-1} A = [I + B(K_0-z)^{-1} A]^{-1}$$

$$= (I + W_z)^{-1} \quad .$$

This implies

$$I - A(H-\lambda\pm i\varepsilon)^{-1} B = [I - B(H-\lambda\mp i\varepsilon)^{-1} A]^*$$

$$= ((I + W_{\lambda\pm i\varepsilon})^{-1})^* \ .$$

Thus by using Proposition 8.9 (b) in the last step :

$$(f, \Omega_\pm E^0_\Delta g) = \lim_{\varepsilon\to+0} \int_\Delta (A^{-1} f, [I - A(H-\lambda\pm i\varepsilon)^{-1}B]\, A\, dE^0_\lambda g)$$

$$= \lim_{\varepsilon\to+0} \int_\Delta ((I + W_{\lambda\pm i\varepsilon})^{-1} A^{-1} f, A\, dE^0_\lambda g)$$

$$= \lim_{\varepsilon\to+0} \int_\Delta ((I + W_{\lambda\pm i\varepsilon})^{-1} A^{-1} f, M_A(\lambda)^*(\mathcal{U}_0 g)_\lambda)\, d\lambda.$$

Now by Lemma 7.5

$$\|(I + W_{\lambda\pm i\varepsilon})^{-1}\| \le M < \infty, \quad \forall\lambda \in \Delta \text{ and } 0 \le \varepsilon \le 1.$$

Also $\|M_A(\lambda)^* (\mathcal{U}_0\, g)_\lambda\| \le \|M_A(\lambda)\|\ \|(\mathcal{U}_0\, g)_\lambda\| \le M_1 < \infty$ for all $\lambda \in \Delta$, since $g \in \mathcal{S}(\mathbb{R}^n)$ and $M_A(\lambda)$ is locally Hölder continuous, hence bounded. Thus the integrand is majorized, uniformly in $0 \le \varepsilon \le 1$, by a constant. Hence by the Lebesgue dominated convergence theorem

$$(f, \Omega_\pm E^o_\Delta g) = \int_\Delta ((I + W_{\lambda\pm io})^{-1} A^{-1} f, M_A(\lambda)^* (\mathcal{U}_0 g)_\lambda)\, d\lambda$$

$$= \int_\Delta (M_A(\lambda)(I + W_{\lambda\pm io})^{-1} A^{-1} f, (\mathcal{U}_0\, g)_\lambda)\, d\lambda. \qquad (11.4)$$

On the other hand

$$(f, \Omega_\pm E^o_\Delta g) = (\Omega^*_\pm f, E^o_\Delta g) = \int_\Delta d\lambda((\mathcal{U}_0\, \Omega^*_\pm f)_\lambda, (\mathcal{U}_0\, g)_\lambda)$$

$$= \int_\Delta d\lambda((\mathcal{U}_\pm f)_\lambda, (\mathcal{U}_0\, g)_\lambda)\ . \qquad (11.5)$$

Since (11.4) and (11.5) also hold for each compact subinterval Δ' of Δ, the integrands must be equal almost everywhere on Δ : There is a null set $N_\pm(f\,, g)$ in Δ (depending on f and g!) such that

$$(M_A(\lambda)(I + W_{\lambda\pm io})^{-1} A^{-1} f, (\mathcal{U}_0\, g)_\lambda)$$
$$= ((\mathcal{U}_\pm f)_\lambda, (\mathcal{U}_0\, g)_\lambda) \quad \text{for all } \lambda \in \Delta\backslash N_\pm(f\,, g)\ . \qquad (11.6)$$

Let $\mathcal{S}_0$ be a countable subset of $\mathcal{S}(\mathbb{R}^n)$ such that for each $\lambda > 0$, the set of vectors $\{(\mathcal{U}_0\, g)_\lambda \mid g \in \mathcal{S}_0\}$ is dense in $L^2(S^{(n-1)})$. ($\mathcal{S}_0$ is the set of all linear combinations with rational coefficients of vectors of the form $\tilde{g}(k) = e^{-|k|^2}\ Y_{\ell m}(\omega)$, where $Y_{\ell m}$ are the surface spherical harmonics).

Let $N_\pm(f) = \bigcup_{g \in S_0} N_\pm(f, g)$. Since S_0 is countable, $N_\pm(f)$ is a null set, and (11.6) holds for all $\lambda \in \Delta \backslash N_\pm(f)$ and all $g \in S_0$. Since $\{(\mathcal{U}_0 g)_\lambda \mid g \in S_0\}$ is dense in $L^2(S^{(n-1)})$:

$$M_A(\lambda)(I + W_{\lambda \pm io})^{-1} A^{-1} f = (\mathcal{U}_\pm f)_\lambda , \quad \forall \lambda \in \Delta \backslash N_\pm(f) . \qquad (11.7)$$

$(\mathcal{U}_\pm f)_\lambda$ is defined only almost everywhere, as a function of λ. We may therefore set

$$(\mathcal{U}_\pm f)_\lambda = M_A(\lambda)(I + W_{\lambda \pm io})^{-1} A^{-1} f \quad \text{for } \textit{all} \ \lambda \in \Delta,$$

which makes $(\mathcal{U}_\pm f)_\lambda$ a strongly continuous function of λ.

Finally, $(0,\infty)\backslash\Gamma_0$ is a countable union of disjoint open intervals. Each $\lambda \in (0,\infty)\backslash\Gamma_0$ belongs to a compact interval $\Delta \subset (0,\infty)\backslash\Gamma_0$, i.e. (11.7) then holds for each $\lambda \in (0,\infty)\backslash\Gamma_0$ (see the proof of Proposition 7.6). ∎

Our next task is to find the eigenfunctions of H. Let us first proceed formally. Since $H\,\Omega_\pm = \Omega_\pm K_0$, we should obtain an eigenfunction $\psi_k^\pm$ of H by acting with $\Omega_\pm$ on the eigenfunction ψ_k^0 of K_0. In view of Proposition 6.5 :

$$\Omega_\pm = I - \underset{\varepsilon \to 0}{\text{s-lim}} \int (K_0 - \lambda \pm i\varepsilon)^{-1} V \,\Omega_\pm \, dE_\lambda^0 \quad ,$$

so that, by acting on ψ_k^0, we expect that $\psi_k^\pm$ will satisfy the following Lippmann-Schwinger equation :

$$\psi_k^\pm = \psi_k^0 - (K_0 - \lambda \pm io)^{-1} V \psi_k^\pm , \text{ with } \lambda = |k|^2 . \qquad (11.8)$$

Now $(K_0 - z)^{-1}$ is an integral operator in $L^2(\mathbb{R}^n)$, namely the operator of convolution with the inverse Fourier transform G_z of the function $g_z(k) = (2\pi)^{-\frac{n}{2}} (k^2 - z)^{-1}$:

$$[(K_0 - z)^{-1} f](x) = \int dy\ G_z(x-y)\ f(y).$$

In particular :

$$[(K_0 - \lambda \mp io)^{-1} f](x) = \int dy\ G_\lambda^{\pm}(x-y)\ f(y).$$

The so-called *Green's functions* (or resolvent kernels) G_z and $G_\lambda^{\pm} = G_{\lambda \pm io}$ are given in terms of Hankel functions as follows :

$$G_z(u) = \frac{i}{4} \left(\frac{\xi}{2\pi|u|}\right)^{\frac{n}{2}-1} H^{(1)}_{\frac{n}{2}-1} (\xi|u|) ,$$

where ξ is the square root of z with Im $\xi > 0$ if $z \notin (0,\infty)$, and $\xi = \pm|\lambda^{\frac{1}{2}}|$ if $z = \lambda \pm io$ with $\lambda > 0$. (Alsholm and Schmidt, [4]), (Titchmarsh, [T]).

Inserting these expressions into (11.8), we obtain the following integral equation for $\psi_k^{\pm}$:

$$\psi_k^{\pm}(x) = \psi_k^{o}(x) - \int dy\ G_\lambda^{\mp}(x-y)\ V(y)\ \psi_k^{\pm}(y). \tag{11.9}$$

One will solve this equation in a suitable function space which must contain $\psi_k^o(x) = (2\pi)^{-\frac{n}{2}} \exp(ik.x)$. The following two spaces have often been used in the literature :

(a) $$L^{\infty}(\mathbb{R}^n) = \{f : \mathbb{R}^n \to \mathbb{C} \mid \operatorname*{ess\,sup}_{x \in \mathbb{R}^n} |f(x)| < \infty\} .$$

(b) $$L^{2,\theta}(\mathbb{R}^n) = \{f : \mathbb{R}^n \to \mathbb{C} \mid \int dx\ (1+|x|)^{\theta}|f(x)|^2 < \infty.$$

Clearly $\psi_k^0 \in L^{\infty}(\mathbb{R}^n)$ and $\psi_k^0 \in L^{2,\theta}(\mathbb{R}^n)$ for each $\theta < -n$. We shall work in $L^{2,\theta}$ and choose $\theta = -2\alpha$, $\alpha > \frac{n}{2}$. We set

$$\mathcal{E} = L^{2,2\alpha}(\mathbb{R}^n) \quad \text{and} \quad \mathcal{E}' = L^{2,-2\alpha}(\mathbb{R}^n) .$$

$\mathcal{E}$ and $\mathcal{E}'$ are Hilbert spaces with scalar products

$$(f , g)_{\mathcal{E}} = \int dx\ (1+|x|)^{2\alpha}\ \overline{f(x)}\ g(x)$$

$$(f , g)_{\mathcal{E}'} = \int dx\ (1 + |x|)^{-2\alpha}\ \overline{f(x)}\ g(x) ,$$

and $\mathcal{E}'$ is the dual space of $\mathcal{E}$: If $f \in \mathcal{E}'$, $g \in \mathcal{E}$, the value of f at g is given by

$$<f , g> = \int dx\ \overline{f(x)}\ g(x) .$$

In the remainder of this chapter we assume the following factorization of V :

$$V(x) = A(x)\ B(x) = (1 + |x|)^{-\alpha}\ [B_1(x) + B_2(x)] \qquad (11.10)$$

with $A(x) = (1 + |x|)^{-\alpha}$ $(\alpha > \frac{n}{2})$, $B_1(.) \in L^{p_1}(\mathbb{R}^n) \cap L^{p_2}(\mathbb{R}^n)$, $p_1 < n < p_2$ and $|B_2(x)| \leq c(1 + |x|)^{-\frac{1}{2}-\nu}$ $(\nu > 0)$.

Multiplication by $A(x) = (1 + |x|)^{-\alpha}$ maps $\mathcal{E}'$ one-to-one onto $L^2(\mathbb{R}^n)$ and $L^2(\mathbb{R}^n)$ one-to-one onto $\mathcal{E}$. Thus $A\,\psi_k^o \equiv \phi_k^o$ is a vector in $L^2(\mathbb{R}^n)$. Here

$$\phi_k^o(x) = (1 + |x|)^{-\alpha} (2\pi)^{-\frac{n}{2}} e^{ik.x} .$$

Setting $\phi_k^{\pm}(x) = A(x)\,\psi_k^{\pm}(x)$, and multiplying (11.8) by A, one obtains

$$\phi_k^{\pm} = \phi_k^o - A(K_o - \lambda \pm io)^{-1} B\, \phi_k^{\pm}$$

$$= \phi_k^o - W^*_{\lambda \pm io}\, \phi_k^{\pm} .$$

If $\lambda \notin \Gamma_o$, this equation has a unique solution in $L^2(\mathbb{R}^n)$, namely

$$\phi_k^{\pm} = ((I + W_{\lambda \pm io})^{-1})^*\, \phi_k^o \quad , \quad (\lambda = |k|^2) . \tag{11.11}$$

Hence $\psi_k^{\pm}(x) \equiv (1 + |x|)^{\alpha}\, \phi_k^{\pm}(x)$ should be two sets of eigenfunctions of H. We first relate these functions to the spectral transformations $\mathcal{U}_{\pm}$ of H.

<u>PROPOSITION 11.2</u> : Let $\mathcal{H} = L^2(\mathbb{R}^n)$, and let V satisfy (11.10). For $\lambda \in (0,\infty)\backslash\Gamma_o$, define

$$\phi_k^{\pm} = ((I + W_{\lambda \pm io})^{-1})^*\, \phi_k^o \quad , \quad (\lambda = |k|^2)$$

and $\psi_k^{\pm} = A^{-1}\, \phi_k^{\pm} \in \mathcal{E}'$. Then, for each $f \in \mathcal{E}$:

(a) $\quad (\mathcal{U}_{\pm} f)_{\lambda}(\omega) = 2^{-\frac{1}{2}} \lambda^{\frac{(n-2)}{4}} \langle \psi_k^{\pm}, f \rangle \quad$, with $k = \sqrt{\lambda}\omega$.

(b) $$\int dk \, |\langle\psi_k^{\pm}, f\rangle|^2 = \|E_{ac}(H)\, f\|^2 .$$

Remarks :

(i) This clearly generalizes the Fourier expansion. In fact, if $V = 0 : \psi_k^{\pm} = \psi_k^o$, i.e.

$$\langle\psi_k^o, f\rangle = (2\pi)^{-\frac{n}{2}} \int dx \, e^{-ik.x} f(x) = \tilde{f}(k) .$$

[Notice that $\mathcal{E} = D(A^{-1}) \in L^1(\mathbb{R}^n) \cap L^2(\mathbb{R}^n)$, and (b) simply reads

$$\int dk \, |\tilde{f}(k)|^2 = \|E_{ac}(K_o)\, f\|^2 = \|f\|^2 .]$$

(ii) The integrand in (b) is not defined if $|k|^2 \in \Gamma_o$. Since this a null set in $\mathbb{R}^n$, this causes no problem. The measurability of the integrand is easy to check. It is also a direct consequence of (a).

Proof : By (8.2), $M_A(\lambda)$ is an integral operator with kernel

$$2^{-\frac{1}{2}} \lambda^{\frac{(n-2)}{4}} (2\pi)^{-\frac{n}{2}} e^{-i\sqrt{\lambda}\omega.x} (1 + |x|)^{-\alpha}$$

$$= 2^{-\frac{1}{2}} \lambda^{\frac{(n-2)}{4}} \overline{\varphi_k^o(x)} ,$$

where $k = \sqrt{\lambda}\omega$. Hence by Proposition 11.1 and (11.11), we obtain for $f \in D(A^{-1}) = \mathcal{E}$:

$$(\mathcal{U}_{\pm} f)_{\lambda} (\omega) = [M_A(\lambda)(I + W_{\lambda\pm io})^{-1} A^{-1} f] (\omega)$$

$$= 2^{-\frac{1}{2}} \lambda^{\frac{(n-2)}{4}} \int dx \overline{\phi_k^0(x)} [(I + W_{\lambda\pm io})^{-1} A^{-1} f](x)$$

$$= 2^{-\frac{1}{2}} \lambda^{\frac{(n-2)}{4}} (\phi_k^0 , (I + W_{\lambda\pm io})^{-1} A^{-1} f)$$

$$= 2^{-\frac{1}{2}} \lambda^{\frac{(n-2)}{4}} ([(I + W_{\lambda\pm io})^{-1}]^* \phi_k^0 , A^{-1} f)$$

$$= 2^{-\frac{1}{2}} \lambda^{\frac{(n-2)}{4}} (\phi_k^{\pm} , A^{-1} f) = 2^{-\frac{1}{2}} \lambda^{\frac{(n-2)}{4}} \langle\psi_k^{\pm} , f\rangle ,$$

which proves (a).

(b) By (a), we obtain upon setting $k^2 = \lambda$

$$\int_{\mathbb{R}^n} dk \, |\langle\psi_k^{\pm} , f\rangle|^2 = 2 \int_0^{\infty} dk \, k^{n-1} \lambda^{-\frac{(n-2)}{2}} \int_{S^{(n-1)}} d\omega \, |(\mathcal{U}_{\pm} f)_{\lambda} (\omega)|^2$$

$$= \int_0^{\infty} d\lambda \int_{S^{(n-1)}} d\omega \, |(\mathcal{U}_o \, \Omega_{\pm}^* f)_{\lambda} (\omega)|^2$$

$$= \|\Omega_{\pm}^* f\|^2 = \|\Omega_{\pm} \Omega_{\pm}^* f\|^2 = \|E_{ac}(H) f\|^2 . \blacksquare$$

<u>PROPOSITION 11.3</u> : Let V satisfy (11.10) and let Δ be a compact subset of $(0,\infty)$ such that $\Delta^2 \cap \Gamma_o = \emptyset$, where $\Delta^2 = \{\alpha^2 \mid \alpha \in \Delta\}$. Then

(a) $\|\psi_k^{\pm}\|_{\mathcal{E}'} \le M_{\Delta} < \infty$ for all $k \in \Delta \times S^{(n-1)}$.

(b) $k \mapsto \psi_k^{\pm}$ is continuous in $\mathcal{E}'$-norm on $\Delta \times S^{(n-1)}$.

Proof :

(a) Let $k \in \Delta \times S^{(n-1)}$. Then by (11.11) :

$$\|\psi_k^{\pm}\|_{\mathcal{E}'} = \|\phi_k^{\pm}\|_2 = \|((I + W_{\lambda \pm io})^{-1})^* \phi_k^0\|_2 \quad (\lambda = k^2)$$

$$\leq \|(I + W_{\lambda \pm io})^{-1}\| \ \|\phi_k^0\|_2 \ .$$

Now $\|(I + W_{\lambda \pm io})^{-1}\| \leq M_\Delta < \infty$ for all $\lambda \in \Delta^2$ by Lemma 7.5, and $\|\phi_k^0\|_2^2 = (2\pi)^{-n} \int dx\ (1 + |x|)^{-2\alpha}\ |e^{ik.x}|^2$ is finite and independent of k. This proves (a).

(b) Let $k, k' \in \Delta \times S^{(n-1)}$. Then, setting $\lambda = k^2$, $\lambda' = k'^2$:

$$\|\psi_k^{\pm} - \psi_{k'}^{\pm}\|_{\mathcal{E}'} = \|\phi_k^{\pm} - \phi_{k'}^{\pm}\|_2$$

$$= \|((I + W_{\lambda \pm io})^{-1})^* (\phi_k^0 - \phi_{k'}^0)$$

$$+ [((I + W_{\lambda \pm io})^{-1})^* - ((I + W_{\lambda' \pm io})^{-1})^*]\ \phi_{k'}^0\|$$

$$\leq M_\Delta \|\phi_k^0 - \phi_{k'}^0\| + \|(I + W_{\lambda \pm io})^{-1} - (I + W_{\lambda' \pm io})^{-1}\|\ \|\phi_{k'}^0\| \ .$$

By Lemma 7.5, $\|(I + W_{\lambda \pm io})^{-1} - (I + W_{\lambda' \pm io})^{-1}\| \to 0$ as $k' \to k$ in $\Delta \times S^{(n-1)}$, and $\|\phi_{k'}^0\|$ is a constant as above. Also

$$\|\phi_k^0 - \phi_{k'}^0\|^2 = (2\pi)^{-n} \int dx\ (1 + |x|)^{-2\alpha}\ |e^{ik.x} - e^{ik'.x}|^2 \to 0$$

as $k' \to k$ by the Lebesgue dominated convergence theorem. Hence $\|\psi_k^{\pm} - \psi_{k'}^{\pm}\|_{\mathcal{E}'} \to 0$ as $k' \to k$. ∎

PROPOSITION 11.4 : $\psi_k^{\pm}$ are eigenfunctions of H in the sense that they are weak solutions of

$$[-\Delta + V(x) - |k|^2]\ \psi(x) = 0 \quad . \quad \text{More precisely :}$$

$$<\psi_k^{\pm}, (H - \lambda)\ f> = 0 \quad \text{for each} \quad f \in \mathcal{S}(\mathbb{R}^n), \ \text{with}\ \lambda = |k|^2 \ .$$

Proof :

(i) Notice that $\mathcal{S}(\mathbb{R}^n) \subseteq \mathcal{E}$ and that $-\Delta f \in \mathcal{E}$, $Vf \in \mathcal{E}$, hence $Hf \in \mathcal{E}$ for each $f \in \mathcal{S}(\mathbb{R}^n)$.

(ii) If V = 0 :

$$\begin{aligned} <\psi_k^{0}, (K_o - \lambda) f> &= (\phi_k^{0}, A^{-1}(K_o - \lambda) f) \\ &= (2\pi)^{-\frac{n}{2}} \int dx\ e^{-ik.x}\ [(-\Delta f)(x) - \lambda f(x)] = 0 \end{aligned} \tag{11.12}$$

upon integration by parts, for each $f \in \mathcal{S}(\mathbb{R}^n)$.

(iii) We use (11.11) and (11.12) to obtain that

$$\begin{aligned} <\psi_k^{\pm}, (H - \lambda)\ f> &= (\phi_k^{\pm}, A^{-1}(H - \lambda)\ f) \\ &= (\phi_k^{0}, (I + W_{\lambda \pm io})^{-1}\ A^{-1}(H - \lambda)\ f) \\ &= \lim_{\varepsilon \to +0}\ (\phi_k^{0}, (I + W_{\lambda \pm i\varepsilon})^{-1}\ A^{-1}(H - \lambda)\ f) \\ &= \lim_{\varepsilon \to +0}\ (\phi_k^{0}, [I - B(H - \lambda \mp i\varepsilon)^{-1}\ A]\ A^{-1}(H - \lambda)\ f) \\ &= \lim_{\varepsilon \to +0}\ [(\phi_k^{0}, A^{-1}(K_o - \lambda)\ f + (\phi_k^{0}, A^{-1}\ V f) \\ &\quad - (\phi_k^{0}, B(H - \lambda \mp i\varepsilon)^{-1}\ (H - \lambda)\ f)] \\ &= (\phi_k^{0}, Bf) - \lim_{\varepsilon \to 0}\ [(\phi_k^{0}, Bf) \pm i\varepsilon\ (\phi_k^{0}, B(H - \lambda \mp i\varepsilon)^{-1}\ f] \\ &= \lim_{\varepsilon \to +0}\ (\mp i\varepsilon)\ (\phi_k^{0}, B(H - \lambda \mp i\varepsilon)^{-1}\ A\ A^{-1}\ f)\ . \end{aligned}$$

Now, by (7.22) :

$$\lim_{\varepsilon\to 0} (\phi_k^0, B(H - \lambda\mp i\varepsilon)^{-1} A A^{-1} f)$$

$$= (\phi_k^0, [I - (I + W_{\lambda\pm io})^{-1}] A^{-1} f) ,$$

hence $$\lim_{\varepsilon\to 0} \varepsilon(\phi_k^0, B(H - \lambda\mp i\varepsilon)^{-1} A A^{-1} f) = 0 . \blacksquare$$

LEMMA 11.5 : Let $V = V_1 + V_2$, $V_1 \in L^p(\mathbb{R}^n)$ with $p > \max(2, \frac{n}{2})$, $V_2 \in L^\infty(\mathbb{R}^n)$. Let $a,b \geq 0$ such that $a + b > \frac{n}{2p}$. Then the closure of

$$(K_0 + 1)^{-a} V(K_0 + 1)^{-b} \quad \text{is in} \quad \mathcal{B}(L^2(\mathbb{R}^n)) .$$

Proof : Clearly $(K_0 + 1)^{-a} V_2(K_0 + 1)^{-b} \in \mathcal{B}(L^2(\mathbb{R}^n))$. Also $(K_0 + 1)^{-a} V_1(K_0 + 1)^{-b}$ is densely defined (e.g. on $\mathcal{S}(\mathbb{R}^n)$), since $(K_0 + 1)^{-b}$ maps $\mathcal{S}(\mathbb{R}^n)$ onto $\mathcal{S}(\mathbb{R}^n)$).

We factorize V_1 into $V_1 = WY$, with

$$W(x) = |V_1(x)|^{\frac{a}{a+b}} , \quad Y(x) = |V_1(x)|^{\frac{b}{a+b}} \operatorname{sign} V_1(x) .$$

We have $W(.) \in L^{\frac{p(a+b)}{a}}(\mathbb{R}^n)$, $Y(.) \in L^{\frac{p(a+b)}{b}}(\mathbb{R}^n)$.

Furthermore $k \mapsto (k^2 + 1)^{-a} \in L^{\frac{p(a+b)}{a}}(\mathbb{R}^n)$ and $k \mapsto (k^2 + 1)^{-b} \in L^{\frac{p(a+b)}{b}}(\mathbb{R}^n)$, since $p(a+b) > \frac{n}{2}$. Notice also that $\frac{p(a+b)}{a} > 2$, $\frac{p(a+b)}{b} > 2$. Hence by Proposition 2.8, the closures of $(K_0 + 1)^{-a} W$ and $Y(K_0 + 1)^{-b}$ are in $\mathcal{B}(L^2(\mathbb{R}^n))$. $\blacksquare$

LEMMA 11.6 : Let V satisfy (11.10) and assume that, for each $\rho \in C_0^\infty(\mathbb{R}^n)$, $\rho(.)\ \psi_k^\pm \in D((K_0+1)^b)$ for some $b \geq 0$. Then for each $\rho \in C_0^\infty(\mathbb{R}^n)$, $\rho(.)\ \psi_k^\pm(.) \in D((K_0+1)^{1-a})$, provided that $a \in [0,1]$ and $a + b > \frac{1}{2}$.

Proof : Since $\psi_k^\pm \in \mathcal{E}'$, we have $\psi_k^\pm \in L^2_{Loc}(\mathbb{R}^n)$, i.e. $\rho(.)\ \psi_k^\pm(.) \in L^2(\mathbb{R}^n)$. Let $f \in \mathcal{S}(\mathbb{R}^n)$, and denotes by $\bar{\rho}$ the operator of multiplication by $\overline{\rho(x)}$. Then by (6.28) :

$$(\rho\psi_k^\pm, (K_0+1)\ f) = <\psi_k^\pm,\ \bar{\rho}(K_0+1)\ f>$$

$$= <\psi_k^\pm, (K_0+1)\ \bar{\rho}f> + <\psi_k^\pm, (\Delta\bar{\rho})\ f>$$

$$+ 2i \sum_{j=1}^n <\psi_k^\pm, (\frac{\partial\bar{\rho}}{\partial x_j})\ P_j\ f> .$$

Now, by Proposition 11.4,

$$<\psi_k^\pm, (K_0+1)\ \bar{\rho}f> = <\psi_k^\pm, (H-\lambda)\ \bar{\rho}f>$$

$$+ <\psi_k^\pm, (1+\lambda - V)\ \bar{\rho}f>$$

$$= <\psi_k^\pm, (1+\lambda - V)\ \bar{\rho}f> .$$

Thus

$$(\rho\psi_k^\pm,\ (K_0+1)\ f) = (1+\lambda)(\rho\psi_k^\pm, f) - (\rho\psi_k^\pm, Vf)$$

$$+ ((\Delta\rho)\psi_k^\pm, f) + 2i \sum_{j=1}^n (\frac{\partial\rho}{\partial x_j}\ \psi_k^\pm, P_j f)$$

$$= (g, (K_0+1)^a\ f) , \qquad (11.13)$$

with

$$g = (1+\lambda)(K_0+1)^{-a}\ \rho\psi_k^{\pm} - [\overline{(K_0+1)^{-a}\ V(K_0+1)^{-b}}]\ [(K_0+1)^{b}\ \rho\psi_k^{\pm}\ ]$$

$$+ (K_0+1)^{-a}\ (\Delta\rho)\ \psi_k^{\pm} - 2i\sum_{j=1}^{n} P_j\ (K_0+1)^{-(a+b)}\ (K_0+1)^{b}\ (\frac{\partial\rho}{\partial x_j})\ \psi_k^{\pm}\ .$$

By hypothesis :

$$(K_0+1)^{b}\ \rho\psi_k^{\pm} \in L^2(\mathbb{R}^n)\ ,\ (K_0+1)^{b}\ (\frac{\partial\rho}{\partial x_j})\ \psi_k^{\pm} \in L^2(\mathbb{R}^n)\quad .$$

Since $a + b > \frac{1}{2}$: $[\overline{(K_0+1)^{-a}\ V\ (K_0+1)^{-b}}] \in \mathcal{B}(\mathcal{H})$ by Lemma 11.5 (notice that $p > n$ as a consequence of the hypothesis (11.10)), and

$$\|P_j(K_0+1)^{-(a+b)}\| = \operatorname*{Sup}_{k\in\mathbb{R}^n} \frac{|k_j|}{(\sum_j k_j^2+1)^{a+b}} \leq 1.$$

It follows that $g \in L^2(\mathbb{R}^n)$. Thus, since $1 - a \geq 0$:

$$(\rho\psi_k^{\pm}\ ,(K_0+1)\ f) = (g\ ,(K_0+1)^{a}\ f)$$

$$= ((K_0+1)^{-(1-a)}\ g\ ,(K_0+1)\ f)\quad .$$

Since the set $\{(K_0+1)\ f \mid f \in \mathcal{S}(\mathbb{R}^n)\}$ is dense in $L^2(\mathbb{R}^n)$ [in fact equal to $\mathcal{S}(\mathbb{R}^n)$], we have

$$\rho\psi_k^{\pm} = (K_0+1)^{-(1-a)}\ g\quad ,$$

i.e. $\rho\psi_k^{\pm} \in D((K_0+1)^{1-a})$. ∎

In Proposition 11.4 we proved that $\psi_k^{\pm}$ are eigenfunctions of H in the sense of distributions. This does not mean that $\psi_k^{\pm}(x)$ have to be differentiable functions of x. The next theorem goes somewhat further in that direction : we show that $\psi_k^{\pm}$ are locally in the domain of K_o.

<u>PROPOSITION 11.7</u> : Let V satisfy (11.10). Then for each $\rho \in C_o^{\infty}(\mathbb{R}^n)$, $\rho(.)\ \psi_k^{\pm}(.) \in D(K_o)$, and the functions $k \mapsto (K_o + 1)\ \rho\psi_k^{\pm}$ are strongly continuous on $\mathbb{R}^n\backslash[\{0\} \cup \Gamma_o^{1/2} \times S^{(n-1)}]$.

<u>Proof</u> :

(i) Take in Lemma 11.6 : b = 0, a = $\frac{2}{3}$. It follows that $\rho(.)\ \psi_k^{\pm}(.) \in D((K_o+1)^{1/3})$. Next take b = 1/3, a = 1/3 to obtain that $\rho(.)\ \psi_k^{\pm}(.) \in D((K_o + 1)^{2/3})$. Finally take b = $\frac{2}{3}$, a = 0 to get $\rho(.)\ \psi_k^{\pm}(.) \in D((K_o + 1)) = D(K_o)$.

(ii) By (11.13) we have for $k^2, k'^2 \notin \Gamma_o \cup \{0\}$:

$$(\rho\psi_k^{\pm} - \rho\psi_{k'}^{\pm}, (K_o+1)f) = (g_o, (K_o+1)^a f) \quad ,$$

with

$$g_o = (1 + \lambda)(K_o+1)^{-a}\ \rho\psi_k^{\pm} - (1 + \lambda')(K_o + 1)^{-a}\ \rho\psi_{k'}^{\pm}$$

$$- \overline{[(K_o+1)^{-a}\ V(K_o+1)^{-b}]}\ [(K_o+1)^b\ \rho\psi_k^{\pm} - (K_o+1)^b\ \rho\psi_{k'}^{\pm}]$$

$$+ (K_o+1)^{-a}\ [(\Delta\rho)\ \psi_k^{\pm} - (\Delta\rho)\ \psi_{k'}^{\pm}]$$

$$- 2i \sum_{j=1}^{n} P_j(K_o+1)^{-(a+b)}\ [(K_o+1)^b\ \frac{\partial\rho}{\partial x_j}\ \psi_k^{\pm}$$

$$- (K_o+1)^b\ \frac{\partial\rho}{\partial x_j}\ \psi_{k'}^{\pm}].$$

By Proposition 11.3 (b) : $k \mapsto \rho\psi_k^{\pm}$, $k \mapsto (\Delta\rho)\,\psi_k^{\pm}$ and $k \mapsto \frac{\partial\rho}{\partial x_j}\,\psi_k^{\pm}$ are strongly continuous in $L^2(\mathbb{R}^n)$.

Assume that $(K_0 + 1)^b\, \rho\psi_k^{\pm}$ is strongly continuous for each $\rho \in C_0^{\infty}(\mathbb{R}^n)$. Then

$$\|g_0\| \to 0 \quad \text{as} \quad k' \to k \ .$$

Since $(K_0 + 1)^{1-a}\, \rho\psi_k^{\pm} - (K_0 + 1)^{1-a}\, \rho\psi_{k'}^{\pm} = g_0$ as in Lemma 11.6, it follows that $(K_0 + 1)^{1-a}\, \rho\psi_k^{\pm}$ is a strongly continuous function of k if $a \in [0,1]$, $a + b > \frac{1}{2}$. By starting with $b = 0$, $a = \frac{2}{3}$ and iterating as in (i), one obtains that $k \mapsto (K_0 + 1)\, \rho\psi_k^{\pm}$ is strongly continuous. ∎

To conclude this chapter we give some further properties of the eigenfunctions $\psi_k^{\pm}$. Their proof will be based on the integral equation (11.9) and requires the knowledge of the following property of the Green's function $G_z(u)$ (G.N. Watson, [W]).

Let Δ be a compact subset of $(0,\infty)$. Then there is a constant $b_n(\Delta)$ such that for all $\lambda \in \Delta$:

$$|G_{\lambda \pm io}(u)| \le \begin{cases} b_n(\Delta)\,|u|^{-(n-2)} & \text{if } |u| \le 1 \\ b_n(\Delta)\,|u|^{\frac{(1-n)}{2}} & \text{if } |u| \ge 1 \ . \end{cases} \tag{11.14}$$

We shall use the following two inequalities :

<u>Young's inequality</u> :

$$\|f * g\|_r \le \|f\|_s\, \|g\|_q \quad \text{if} \quad s^{-1} + q^{-1} - r^{-1} = 1 \tag{11.15}$$
$$1 \le q\,, r\,, s \le \infty$$

(* denotes the convolution),

Sobolev's inequality :

$$\iint dx\, dy\, |f(x)|\, |g(y)|\, |x-y|^{-\gamma} \le C\|f\|_r \|g\|_q \tag{11.16}$$

if $1 < q,r < \infty$, $\gamma \in (0,n)$ and $r^{-1} + q^{-1} + \gamma n^{-1} = 2$.

Since $L^r(\mathbb{R}^n)$ is the dual space of $L^{r'}(\mathbb{R}^n)$, $r' = (1-r^{-1})^{-1}$, (11.16) implies that

$$x \mapsto h(x) \equiv \int dy\, |x-y|^{-\gamma}\, g(y) \in L^{r'}(\mathbb{R}^n) \tag{11.17}$$

and

$$\|h\|_{r'} \le C\|g\|_q \quad . \tag{11.18}$$

PROPOSITION 11.8 : Let V satisfy (11.10). Let Δ^2 be a compact subset of $(0,\infty)\setminus\Gamma_o$. Then $(x,k) \mapsto \psi_k^{\pm}(x)$ is bounded on $\mathbb{R}^n \times (\Delta \times S^{(n-1)})$. In particular $\psi_k^{\pm}(.) \in L^{\infty}(\mathbb{R}^n)$ for each $k \in \Delta \times S^{(n-1)}$.

Proof : By (11.9)

$$\psi_k^{\pm}(x) = \psi_k^{o}(x) - \sum_{i,j=1}^{2} h_{ji}(x) \tag{11.19}$$

with

$$h_{1,i}(x) = \int_{|x-y|\ge 1} dy\, G_\lambda^{\mp}(x-y)\, B_i(y)\, \phi_k^{\pm}(y) \tag{11.20}$$

$$h_{2,i}(x) = \int_{|x-y|\le 1} dy\, G_\lambda^{\mp}(x-y)\, B_i(y)\, \phi_k^{\pm}(y) \tag{11.21}$$

Clearly ψ_k^0 has all properties stated in the proposition. We have to show the same for $h_{1,i}$ and $h_{2,i}$ $(i = 1,2)$.

(i) We first treat $h_{1,i}$. Setting $b_n'(\Delta) = b_n(\Delta^2)$, we have

$$|h_{1,i}(x)| \le b_n'(\Delta) \int_{|x-y|\ge 1} dy \, |x-y|^{\frac{(1-n)}{2}} \, |B_i(y)| \, |\phi_k^{\pm}(y)|.$$

Applying Young's inequality with $f(u) = \chi_{\mathbb{R}^n \setminus S_1}(u) \, |u|^{\frac{(1-n)}{2}}$,

$g(u) = |B_i(u) \, \phi_k^{\pm}(u)|$ and $r = \infty$, we obtain (a_{n-1} denotes the surface area of $S^{(n-1)}$)

$$\|h_{1,i}\|_\infty \le b_n'(\Delta) \, a_{n-1}^{\frac{1}{s_i}} \left| \int_1^\infty du \, u^{n-1} \, u^{\frac{s_i(1-n)}{2}} \right|^{\frac{1}{s_i}} \|B_i \, \phi_k^{\pm}\|_{q_i} . \qquad (11.22)$$

By Proposition 11.3(a) : $\|\phi_k^{\pm}\|_2 \le M_\Delta < \infty$ for all $k \in \Delta \times S^{(n-1)}$. Furthermore

$$B_1(.) \in L^n(\mathbb{R}^n) \quad , \quad B_2(.) \in L^{2n-\varepsilon}(\mathbb{R}^n) \quad \text{for some } \varepsilon > 0.$$

By the Hölder inequality (2.10) :

$$B_i \, \phi_k^{\pm} \in L^{q_i}(\mathbb{R}^n) \quad , \quad q_1 = \frac{2n}{(n+2)} \quad , \quad q_2 = \frac{2n}{(n+1)} - \varepsilon' \quad , \qquad (11.23)$$

for some $\varepsilon' > 0$. By (11.15) we must have $s_i^{-1} + q_i^{-1} = 1$, hence $s_1 = \frac{2n}{(n-2)}$ and $s_2 = \frac{2n+\varepsilon''}{(n-1)}$ for some $\varepsilon'' > 0$. In both cases the integral over du in (11.22) is finite, so that we have shown that $h_{1,i} \in L^\infty(\mathbb{R}^n)$ and that $\|h_{1,i}\|_\infty < C(\Delta)$ for all $k \in \Delta \times S^{(n-1)}$.

(ii) We now treat $h_{2,i}$. Suppose we know that $\phi_k^{\pm} \in L^q(\mathbb{R}^n)$ for some $q \geq 2$. Then, as in (11.23) ,

$$B_i \phi_k^{\pm} \in L^{q_i}(\mathbb{R}^n) \quad \text{with} \quad q_1 = \frac{nq}{n+q} \quad , \quad q_2 = \frac{2nq}{2n+q} - \varepsilon. \qquad (11.24)$$

By using Sobolev's inequality with $\gamma = n-2$, we now obtain that $h_{2,i} \in L^{t_i}(\mathbb{R}^n)$ with $t_1 = nq/(n-q)$ and $t_2 = 2nq/(2n-3q) - \varepsilon'''$ ($\varepsilon''' > 0$). Notice that $t_2 > t_1 > q$. This makes sense provided $q \leq \frac{2n}{3}$.

Now by (11.19) :

$$\phi_k^{\pm} \equiv A\psi_k^{\pm} = A(\psi_k^0 - h_{11} - h_{12}) - A(h_{21} + h_{22}) \ . \qquad (11.25)$$

We have $A \in L^p(\mathbb{R}^n)$ for each $p \geq 2$. By (i), the first term on the r.h.s. is in $L^p(\mathbb{R}^n)$ for each $p \in [2,\infty]$. Furthermore $Ah_{21} \in L^{t_1}(\mathbb{R}^n)$, and also $Ah_{22} \in L^{t_1}(\mathbb{R}^n)$ by applying the Hölder inequality. Hence $\phi_k^{\pm} \in L^q(\mathbb{R}^n)$ implies that $\phi_k^{\pm} \in L^{t_1}(\mathbb{R}^n)$.

Since $\phi_k^{\pm} \in L^2(\mathbb{R}^n)$, this implies (take $q = 2$) that $\phi_k^{\pm} \in L^{2n/(n-2)}(\mathbb{R}^n)$. Next take $q \in [2,2n/(n-2)]$ to obtain that $\phi_k^{\pm} \in L^t(\mathbb{R}^n)$ for each $t \in [2,2n/(n-4)]$. Iterating, with $q \leq 2n/3$, one finds that $\phi_k^{\pm} \in L^t(\mathbb{R}^n)$ for each $t \in [2,2n)$. Hence $B\phi_k^{\pm} \in L^n(\mathbb{R}^n)$. Finally we apply Young's inequality (11.15) with $r = \infty$, $q = n$ in the following integral :

$$|h_2(x)| \leq b_n(\Delta) \int_{|x-y|\leq 1} dy \ |x-y|^{-n+2} \ |B(y) \ \phi_k^{\pm}(y)|$$

$$\leq b_n'(\Delta) \ \|B \ \phi_k^{\pm}\|_n \left[a_{n-1} \int_0^1 du \ u^{n-1} \ u^{-n(n-2)/(n-1)}\right]^{\frac{n-1}{n}} < \infty.$$

This shows that $h_2 \in L^\infty(\mathbb{R}^n)$. It is also easily checked that all estimates above are uniform in $k \in \Delta \times S^{(n-1)}$. ■

PROPOSITION 11.9 : Let V satisfy (11.10), and let $k \in \mathbb{R}^n$ be such that $k^2 \notin \Gamma_0 \cup \{0\}$. Then

$$|\psi_k^{\pm}(x) - \psi_k^0(x)| = O(|x|^{-\kappa}) \quad \text{as} \quad |x| \to \infty, \tag{11.26}$$

with $\kappa = \min\left(\frac{n-1}{2}, \nu\right)$, where ν is the number appearing in the bound for B_2 in (11.10).

Proof : From (11.9), Proposition 11.8 and (11.14) we find that

$$\begin{aligned} |\psi_k^{+}(x) - \psi_k^0| &\le c_k \int \{|x-y|^{-n+2} + |x-y|^{-\frac{(n-1)}{2}}\} |V(y)| \, dy \\ &\le c_k \int \{|x-y|^{-n+2} + |x-y|^{-\frac{(n-1)}{2}}\} (1 + |y|)^{-\frac{n-1}{2} - \nu} W(y) \, dy, \end{aligned} \tag{11.27}$$

where $W(y) = |B_1(y)| + (1 + |y|)^{-1-\delta}$ $(\delta > 0)$, hence $W(.) \in L^n(\mathbb{R}^n)$. By applying the Hölder inequality to (11.27) we obtain

$$|\psi_k^{\pm}(x) - \psi_k^0(x)| \le c_k \|W\|_n (I_1(x) + I_2(x)) \quad ,$$

with

$$I_i(x) = \left[\int |x-y|^{-\alpha_i} (1 + |y|)^{-\frac{n}{2} - \frac{n\nu}{(n-1)}} dy\right]^{\frac{n-1}{n}} , \tag{11.28}$$

$\alpha_1 = \frac{n(n-2)}{(n-1)}$, $\alpha_2 = \frac{n}{2}$.

The result of the Proposition follows from (11.28) and the following Lemma. ∎

<u>LEMMA 11.10</u> : Let $\gamma \in (0,n)$ and $\gamma + \theta > n$. Then as $|x| \to \infty$,

$$\int_{\mathbb{R}^n} |x-y|^{-\gamma} (1+|y|)^{-\theta}\, dy = \begin{cases} O(|x|^{n-\gamma-\theta}) & \text{if } \theta < n \\ O(|x|^{-\gamma}) & \text{if } \theta > n \end{cases}.$$

<u>Proof</u> : Let $|x| > 1$, and divide the domain of integration into $\Delta_> \cup \Delta_<$, where $\Delta_{\gtrless} = \{y \in \mathbb{R}^n \mid |y| \gtrless \frac{1}{2}|x|\}$. In $\Delta_>$, we make the change of variables $z = |x|^{-1} y$ and obtain

$$\int_{\Delta_>} |x-y|^{-\gamma} (1+|y|)^{-\theta}\, dy \le |x|^{n-\gamma-\theta} \int_{|z|\ge\frac{1}{2}} |\hat{x}-z|^{-\gamma} |z|^{-\theta}\, dz ,$$

where $\hat{x} = |x|^{-1} x$. Since $\gamma+\theta > n$, the last integral is finite and is independent of $\hat{x} \in S^{(n-1)}$, hence this term is $O(|x|^{n-\gamma-\theta})$.

For the second integral we have

$$\int_{\Delta_<} |x-y|^{-\gamma}(1+|y|)^{-\theta}\, dy \le \left(\frac{2}{|x|}\right)^{\gamma} \int_{|y|\le\frac{|x|}{2}} (1+|y|)^{-\theta}\, dy ,$$

which is $O(|x|^{-\gamma})$ if $\theta > n$ and $O(|x|^{n-\theta-\gamma})$ if $\theta < n$. ∎

12 Geometrical approach

A. INTRODUCTION AND RESULTS.

When the potential V vanishes at infinity fast enough to exclude the Coulomb case, the completeness of the wave operators $W_{\pm} = s - \lim e^{itH} e^{-itK_0}$ as $t \to \pm\infty$ can be proved using eigenfunction expansions (Agmon, [1]) or by an abstract method.

By using a time-dependent method V. Enss in [19] gave a new proof of the completeness of the wave operators and the non-existence of singular continuous spectrum of H. This method is based on a study of how a scattering particle behaves in space and in time : This gives the "geometrical" or "time-dependent" approach introduced by V. Enss in [19] and developped further by various authors (Amrein, Pearson and Wollenberg [9], Davies [16], Enss [20], Ginibre [22], Mourre [29], Perry [32], Simon [35], Yafaev [38]).

The main idea of the proof of V. Enss is the following : Ruelle [33] and Amrein and Georgescu [6] have shown in two nice papers that, under suitable assumptions on the potential V, each state (represented by a vector, see Chapter 13) from the subspace of continuity $\mathcal{H}_c(H)$ of the Hamiltonian $H = K_0 + V$ leaves in the time mean any finite region of space. When the state is localized far away from the region in which the potential is appreciably different from zero, it may be decomposed into two pieces, one with velocities pointing outward and one with velocities pointing inward (phase-space decomposition of state vectors). The first piece will not be influenced much by the potential in the future, the second one in the past. By using this decomposition of states, one can then show that each vector in $\mathcal{H}_c(H)$ orthogonal to the range of $\Omega_{\pm}$ must be zero, implying that range $\Omega_{\pm} = \mathcal{H}_c(H)$.

The potential V, without symmetry, is assumed to be a relatively bounded perturbation of K_0 with K_0-bound smaller than 1 :

$$\|Vf\| \leq \alpha\|f\| + \beta\|K_0 f\| \ , \ \forall f \in D(K_0) \ , \beta < 1 . \qquad (12.1)$$

The total Hamiltonian H is given by

$$H = K_0 + V .$$

This defines a self-adjoint operator with domain $D(H) = D(K_0) \subset D(V)$. For a subset $\Sigma \subset \mathbb{R}^n$, define the operator F_Σ to be $[F_\Sigma g](x) = \chi_\Sigma(x) g(x)$.

We add also a condition on the potential at infinity :

$$\|V(K_0 + i)^{-1} F_{|x|\geq R}\| \equiv h(R) \in L^1((0,\infty) , dR) , \qquad (12.2)$$

where $F_{|x|\geq R}$ is the operator F_Σ associated with $\Sigma = \{x \mid |x| \geq R\}$.

The principal result is the following :

<u>PROPOSITION 12.1</u> : Assume that $V : \mathbb{R}^n \to \mathbb{R}$ satisfies (12.1) and (12.2). Then the wave operators $\Omega_\pm$ exist, and the range of $\Omega_\pm$ is the subspace of $\mathcal{H}$ corresponding to the continuous spectrum of $\mathcal{H}$: range $\Omega_\pm = \mathcal{H}_c(H)$.

The proof of this theorem will occupy the main part of this chapter. In an appendix we shall give some results on modified wave operators for potentials tending to zero at infinity more slowly than the Coulomb potential (so-called long range potentials), using a similar type of technique.

B. WEAK DECAY RESULTS.

We begin with a result which says that any state in the subspace of continuity $\mathcal{H}_c(H)$ of H decays locally, in time mean ; we claim

LEMMA 12.2 :

$$\lim_{T\to\infty} \frac{1}{T}\int_0^T \|F_{|x|\le R}\, e^{-i\tau H} f\|^2 \, d\tau = 0 \ , \ \forall R < \infty, \tag{12.3}$$

provided that $f \in \mathcal{H}_c(H)$.

Proof : All we need is that

$$F_{|x|\le R}\,(H+i)^{-1} \quad \text{is compact.} \tag{12.4}$$

Indeed, it suffices to show the Lemma for $f \in D(H) \cap \mathcal{H}_c(H)$; say $f = (H+i)^{-1} g$, $g \in \mathcal{H}_c(H)$. We need to prove only that

$$\lim_{T\to\infty} \frac{1}{T}\int_0^T \|K\, e^{-i\tau H} g\|^2 \, d\tau = 0 \tag{12.5}$$

for $g \in \mathcal{H}_c(H)$, K any compact operator. It suffices to show (12.5) holds for all finite rank operators K, and hence it suffices to consider K self-adjoint of rank 1 ; $Kf = (h,f)h$. Thus we need only show that

$$\lim_{T\to\infty} \frac{1}{T}\int_0^T |(h, e^{-i\tau H} g)|^2 \, d\tau = 0 \tag{12.6}$$

for any $h \in L^2(\mathbb{R}^n)$, $g \in \mathcal{H}_c(H)$. Writing $e^{-i\tau H} = \int_{-\infty}^{+\infty} e^{-i\tau\lambda}\, dE_\lambda$, we see that

$$(h, e^{-i\tau H} g) = \int_{-\infty}^{+\infty} e^{-i\tau\lambda} d(h, E_\lambda g) = \tilde{\mu}(\tau)$$

where the finite measure μ on $\mathbb{R}$ is given by $d\mu(\lambda) = d(h, E_\lambda g)$. The hypothesis that $g \in \mathcal{H}_c(H)$ implies that μ has no point masses. The left-side of (12.6) is $\lim \frac{1}{T}\int_0^T |\tilde{\mu}(\tau)|^2 \, d\tau$ as $T \to \infty$, so (12.6) follows from a result of

(Wiener, [37]) , which says that, if μ is a finite measure on the line $\mathbb{R}$, $\phi \in L^1(\mathbb{R})$, $\int_{\mathbb{R}} \phi(\tau)\, d\tau = 1$, then

$$\lim_{T\to\infty} \frac{1}{T} \int_{\mathbb{R}} \phi\,(T^{-1}\tau)\ |\tilde{\mu}(\tau)|^2\, d\tau = \gamma \tag{12.7}$$

where γ is the sum of the squares of the point masses of μ. Indeed,

$$T^{-1} \int \phi(T^{-1}\tau)\,|\tilde{\mu}(\tau)|^2\, d\tau = T^{-1} \iiint \phi(T^{-1}\tau)\, e^{-i\tau\lambda}\, d\mu(\lambda)\, e^{i\tau\sigma}\, d\overline{\mu(\sigma)}\, d\tau$$

$$= \iint \tilde{\phi}(T(\lambda-\sigma))\, d\mu(\lambda)\, d\overline{\mu(\sigma)} \quad .$$

This last quantity tends to $\iint_{\lambda=\sigma} d\mu(\lambda)\, d\overline{\mu(\sigma)} = \gamma$, where we have applied the Lebesgue dominated convergence theorem and the Riemann-Lebesgue Lemma. This proves (12.7). ∎

We will use the above mean decay result to construct from $f \in \mathcal{H}_c(H)$ a sequence $f_\nu = e^{-i\tau_\nu H} f$ which decays locally. Indeed, we suppose $f \in D(H) \cap \mathcal{H}_c(H)$. Setting $R = \nu$ in (12.4), we assert :

<u>LEMMA 12.3</u> : $\int_{-\nu}^{+\nu} \| F_{|x|<\nu}\ e^{-i(t+\tau)H}\ (H+i)f \|\, dt \to 0$ in the mean, as a function of τ, as $\tau \to \infty$, for ν fixed.

<u>Proof</u> : Let $g = (H+i)f$. Then

$$\frac{1}{T}\int_0^T \int_{-\nu}^{+\nu} \| F_{|x|<\nu}\ e^{-i(t+\tau)H}\ g\|\ dt\, d\tau$$

$$= \int_{-\nu}^{+\nu} \left(\frac{1}{T}\int_0^T \| F_{|x|<\nu}\ e^{-i(t+\tau)H} g\|\ d\tau\right) dt \to 0 \text{ as } T \to \infty,$$

by Lemma 12.2, and the Lebesgue dominated convergence theorem, since mean square convergence to zero implies mean convergence to zero. ∎

Using Lemmas 12.2 and 12.3 we can pick a sequence $\{\tau_\nu\}$ such that $f_\nu = e^{-i\tau_\nu H} f$ decays locally. For later purposes, it is necessary to make a stronger, somewhat technical assertion . Let $f \in D(H) \cap \mathcal{H}_c(H)$ be as above

<u>LEMMA 12.4</u> : There exist $a_\nu \uparrow \infty$ and subsets Σ_ν of $I_\nu \equiv (a_\nu, a_{\nu+1})$, with $(a_{\nu+1} - a_\nu)^{-1} \cdot \text{meas } \Sigma_\nu \to 1$ as $\nu \to \infty$, such that, if $\tau_\nu \in \Sigma_\nu$, then as $\nu \to \infty$

$$\| F_{|x|<n\nu^2} \, e^{-i\tau_\nu H} f \| \to 0 \tag{12.8}$$

and

$$\int_{-\nu}^{+\nu} \| F_{|x|<\nu} \, e^{-i(\tau_\nu + t)H} (H+i) f \| \, dt \to 0 \, . \tag{12.9}$$

<u>Proof</u> : Let $\{\phi_k\}$ be a sequence of non-negative functions and assume that $\frac{1}{T}\int_0^T \phi_k(\tau)\, d\tau \to 0$ as $T \to \infty$ for each k. Let $a_1 = 1$ and choose $\{a_\nu\}_{\nu=2}^{\infty}$ recursively as follows. Given $a_1, \ldots, a_\nu$, set $\Gamma_{\nu,s} = \{\tau \in (a_\nu, s) \mid \phi_\nu(\tau) < \frac{1}{\nu}\}$ and $\Gamma'_{\nu,s} = (a_\nu, s) \setminus \Gamma_{\nu,s}$. Since $\phi_\nu(\tau) \geq \nu^{-1}$ for $\tau \in \Gamma'_{\nu,s}$, we have $\chi_{\Gamma'_{\nu,s}}(\tau) \leq \nu\phi_\nu(\tau)$. Hence, if $s > 2a_\nu$ (i.e. $s - a_\nu > \frac{s}{2}$) :

$$(s-a_\nu)^{-1} \text{ meas } \Gamma'_{\nu,s} \leq (s-a_\nu)^{-1} \int_0^s \nu\phi_\nu(\tau)\, d\tau$$

$$\leq \frac{2\nu}{s} \int_0^s \phi_\nu(t) \to 0 \text{ as } s \to \infty.$$

Hence we may choose $s = a_{\nu+1}$ such that $(a_{\nu+1} - a_\nu)^{-1}$ meas $\Gamma_{\nu,s} > 1 - \frac{1}{\nu}$. If we set $\Sigma_\nu = \Gamma_{\nu, a_{\nu+1}}$, we clearly have $\lim \phi_\nu(\tau_\nu) = 0$ as $\nu \to \infty$ if $\tau_\nu \in \Sigma_\nu$.

(12.8) and (12.9) follow from Lemmas 12.2 and 12.3 by taking

$$\phi_\nu(\tau) = \|F_{|x|<n\nu^2}\, e^{-i\tau H} f\|^2 + \int_{-\nu}^{+\nu} \|F_{|x|<\nu}\, e^{-i(t+\tau)H}(H+i)f\|\, dt. \blacksquare$$

For any $\tau_\nu \in \Sigma_\nu$, set

$$f_\nu = e^{-i\tau_\nu H} f \tag{12.10}$$

These states are "small" on $|x|<n\nu^2$. The next lemma asserts that their kinetic energy distributions approximate their total energy distributions.

LEMMA 12.5 : If $\phi \in L^1(\mathbb{R})$, $\tilde{\phi}(s) = (2\pi)^{-\frac{1}{2}} \int e^{-ist}\, \phi(t)\, dt$, then

$$\lim_{\nu \to \infty} \|(\tilde{\phi}(H) - \tilde{\phi}(K_0))f_\nu\| = 0 .$$

Proof : We can characterize $\tilde{\phi}(H) = (2\pi)^{-\frac{1}{2}} \int_{-\infty}^{+\infty} \phi(t)\, e^{-itH}\, dt$, so

$$(2\pi)^{\frac{1}{2}} \|[\tilde{\phi}(H) - \tilde{\phi}(K_0)]f_\nu\| = \left\|\int_{-\infty}^{+\infty} \phi(t)\, (e^{-itH} - e^{-itK_0})\, f_\nu\, dt\right\|$$

$$\leq \int_{-\nu}^{+\nu} |\phi(t)|\, \|(e^{itK_0}\, e^{-itH} - I)\, f_\nu\|\, dt +$$

$$+ 2\,\|f\| \int_{|t|>\nu} |\phi(t)|\, dt .$$

The second term clearly vanishes as $\nu \to \infty$. The first term is bounded by

$$\|\phi\|_1 \sup_{|t|\le\nu} \|(e^{itK_0} e^{-itH} - I)f_\nu\| \le \|\phi\|_1 \int_{-\nu}^{\nu} \|V e^{-itH} f_\nu\| \, dt$$

$$\left.\begin{aligned} &\le \|\phi\|_1 \int_{-\nu}^{+\nu} \|V(H+i)^{-1}\| \, \|F_{|x|<\nu} \, e^{itH}(H+i)f_\nu\| \, dt \\ &+ \|\phi\|_1 \int_{-\nu}^{+\nu} \|V(H+i)^{-1} F_{|x|\ge\nu}\| \, dt \, \|(H+i)f\| . \end{aligned}\right\} \qquad (12.11)$$

The first term in (12.11) vanishes as $\nu \to \infty$, by (12.9), and the second term vanishes because of our short range hypothesis (12.2) ; indeed (12.2), the second resolvent equation (4.8) and the fact that V is H-bounded imply that

$$\|V(H+i)^{-1} F_{|x|\ge\nu}\| \le c \, h(\nu) ,$$

which is in $L^1(\mathbb{R})$ and monotone decreasing as a function of ν, so that $\nu h(\nu) \to 0$ as $\nu \to \infty$. This proves the lemma. ∎

C. ASYMPTOTIC DECOMPOSITION OF STATES.

Let $\{f_\nu\}$ be defined as in (12.10). We will assume that the vector f is in a dense subset M of $\mathcal{H}_c(H)$ which contains all states whose energy is finite and bounded away from zero (i.e. $E_{(\alpha,\beta)} \mathcal{H}_c(H) \subset M$ for each $0 < \alpha < \beta < \infty$, where $\{E_\lambda\}$ is the spectral family of H).

Let Δ be a finite interval in $(0,\infty)$ such that $E_\Delta f = f$. Choose a function $\phi \in \mathcal{S}(\mathbb{R})$ such that $\tilde{\phi}(\lambda) = 1$ if $\lambda \in \Delta$ and such that $\tilde{\phi}$ has compact support in $(0,\infty)$. Thus there exist two numbers $a,b \in (0,\infty)$ such that the support of $\tilde{\phi}$ is contained in the interval $(16a^2,b^2)$. We clearly have $f_\nu = \tilde{\phi}(H)f_\nu$.

Let us also define

$$g_\nu = \tilde{\phi}(K_0) \, f_\nu \, . \qquad (12.12)$$

Notice that g_ν has momentum support in the set $4a \le |k| \le b$, i.e. $\tilde{g}_\nu(k) = 0$ if k does not belong to this set. Lemma 12.5 implies that

$$\lim_{\nu \to \infty} \| f_\nu - g_\nu \| = 0 . \tag{12.13}$$

We next introduce for each ν a decomposition of the space $\mathbb{R}^n$ into a cube B_ν and finitely many truncated cones $C_{\nu,j}$. For this we set

$$B_\nu = \{x \in \mathbb{R}^n \mid |x_i| \le \nu^2 - \frac{1}{2} \quad \text{for all} \quad i = 1,\dots,n\}$$

and

$$C_j = \{x \in \mathbb{R}^n \mid x.e_j \ge \frac{99}{100} |x|\} \quad , \quad j = 1,\dots,m .$$

Here $e_1,\dots,e_m$ are a finite number of fixed unit vectors (independent of ν) such that $\bigcup_j C_j = \mathbb{R}^n$. For each j and ν we pick a subset $C_{\nu,j}$ of C_j such that

$$\mathbb{R}^n = B_\nu \cup C_{\nu,1} \cup \dots \cup C_{\nu,m} \tag{12.14}$$

and such that the m+1 sets appearing in (12.14) are pairwise disjoint, thus obtaining the decomposition mentioned above.

It will be necessary to consider an even finer decomposition of $\mathbb{R}^n$. For this, we denote by Λ_μ the unit cube centered at $\mu = (\mu_1,\dots,\mu_n)$, where $\mu_1,\dots,\mu_n$ are integers. We write $\mu \div (\nu,j)$ if the point μ lies in the truncated cone $C_{\nu,j}$ and then have the decomposition

$$\mathbb{R}^n = B_\nu \cup (\bigcup_j \bigcup_{\mu \div (\nu,j)} \Lambda_\mu) . \tag{12.15}$$

In (12.15) we have approximately (i.e. up to insignificant boundary effects) decomposed the truncated cone $C_{\nu,j}$ into a union of cubes $\bigcup_{\mu\div(\nu,j)} \Lambda_\mu$.

We next fix a function η in $S(\mathbb{R}^n)$ such that $\tilde{\eta}(k)$ has support on $|k| \leq a$ and such that $\int \eta(x)\,dx = 1$. If χ_Σ is the characteristic function of the set $\Sigma \subset \mathbb{R}^n$, we denote by $G(\Sigma)$ the multiplication operator by the function $\eta * \chi_\Sigma$ (* means convolution). (12.15) then leads to the following decomposition of the identity operator I :

$$I = G(B_\nu) + \sum_{j=1}^{m} \sum_{\mu\div(\nu,j)} G(\Lambda_\mu) . \tag{12.16}$$

Each vector $G(\Lambda_\mu)g_\mu$ is essentially localized far from the scatterer, for large ν, and will be decomposed further into its "outgoing" and "incoming" components. For this, we introduce for each j a function η_j which belongs to $C_0^\infty(\mathbb{R}^n)$ and is such that

$$\eta_j(k) + \eta_j(-k) = 1 \quad \text{if} \quad |k| \in [4a,b] \tag{12.17}$$

and such that

$$\eta_j(k) = 0 \quad \text{if} \quad k.e_j \leq -a . \tag{12.18}$$

We then have

$$[\eta_j(k) + \eta_j(-k)]\ \tilde{g}_\nu(k) = \tilde{g}_\nu(k) ,$$

or equivalently

$$[\eta_j(P) + \eta_j(-P)]\ g_\nu = g_\nu , \tag{12.19}$$

where P is the n-component momentum operator.

We now define

$$g_\nu(j\,,\text{out/in}) = \sum_{\mu \div (\nu,j)} G(\Lambda_\mu)\ \eta_j(\pm P) g_\nu \tag{12.20}$$

and

$$g_\nu(\text{out/in}) = \sum_{j=1}^{m} g_\nu(j\,,\text{out/in})\,. \tag{12.21}$$

The following properties of these vectors will be important in proving completeness of the wave operators :

LEMMA 12.6 : The following relations hold :

$$g_\nu = G(B_\nu) g_\nu + g_\nu(\text{out}) + g_\nu(\text{in})\ , \tag{12.22}$$

$$\lim_{\nu\to\infty} \|G(B_\nu) g_\nu\| = 0\ , \tag{12.23}$$

$$\lim_{\nu\to\infty} \|f_\nu - g_\nu(\text{out}) - g_\nu(\text{in})\| = 0\ . \tag{12.24}$$

Proof : (12.22) is immediate from (12.19) and (12.16). (12.24) follows from (12.22), (12.23) and (12.13).

To prove (12.23), we notice that

$$|(\eta * \chi_{B_\nu})(x)| \le \int |\eta(x-y)|\ dy = \|\eta\|_1$$

and that for $x \notin B_\nu$, $|(\eta * \chi_{B_\nu})(x)| \le c[\text{dist}(x,B_\nu)]^{-1}$, since $\eta \in \mathcal{S}(\mathbb{R}^n)$. Using the fact that the distance between B_ν and the set $\{x \mid |x| \ge n\nu^2\}$ is proportional to ν^2, we then get

$$\|G(B_\nu)g_\nu\| \le \|G(B_\nu)(g_\nu - f_\nu)\| + \|G(B_\nu)F_{|x|<n\nu^2} f_\nu\|$$

$$+ \|G(B_\nu)\, F_{|x|\ge n\nu^2}\, f_\nu\|$$

$$\le \|\eta\|_1 \,[\, \|g_\nu - f_\nu\| + \|F_{|x|<n\nu^2}\, f_\nu\|\,] + c_n\, \nu^{-2}\, \|f\| ,$$

which converges to zero as $\nu \to \infty$ by (12.13) and (12.8). ∎

D. LOCAL DECAY OF $e^{-iK_0 t} g_\nu(j\,, \text{out/in})$.

The purpose of this section is to prove the following strong result on the local decay of $e^{-iK_0 t} g_\nu(j\,, \text{out/in})$.

PROPOSITION 12.7 : Under the hypothesis on $g_\nu(j\,, \text{out/in})$ given in part C, one has for each $\varepsilon \le \min(1, \frac{a}{2})$:

$$\lim_{\nu\to\infty} \int_0^\infty \|F_{|x|\le\varepsilon(\nu+t)}\, e^{-iK_0 t} (K_0 + i)\, g_\nu(j\,, \text{out})\|\, dt = 0 \,, \qquad (12.25)$$

and

$$\lim_{\nu\to\infty} \int_{-\infty}^0 F_{|x|\le\varepsilon(\nu-t)}\, e^{-iK_0 t} (K_0 + i)\, g_\nu(j\,, \text{in})\|\, dt = 0 \,. \qquad (12.26)$$

In order to prove such decay, it will be useful to make a general study of the decay of $e^{-iK_0 t}h$ when $\tilde{h}(k)$ has support in some specified set. We accomplish this by a certain decomposition which we now describe. Using (6.21), we may write

$$[e^{-iK_o t} h](x) = (4\pi it)^{-\frac{n}{2}} e^{ix^2/4t} \int e^{iy^2/4t} e^{-i\frac{x.y}{2t}} h(y)\, dy$$

$$= (4\pi it)^{-\frac{n}{2}} e^{ix^2/4t} \int [M_k(\frac{y^2}{4t}) + R_k(\frac{y^2}{4t})]\, e^{-i\frac{x.y}{2t}} h(y)\, dy\,, \tag{12.27}$$

where

$$M_k(s) = \sum_{j=0}^{k} \frac{(is)^j}{j!}\,. \tag{12.28}$$

Note that $|R_k(s)| \le c_k\, |s|^{k+1}$; more generally,

$$\left|\frac{d^\ell}{ds^\ell} R_k(s)\right| \le c|s|^{k+1-\ell}\,, \quad 0 \le \ell \le k+1\,. \tag{12.29}$$

From (12.27) we obtain

$$e^{-itK_o} h = A_k(t)h + B_k(t)h\,, \tag{12.30}$$

where

$$[A_k(t)h]\,(x) = (4\pi it)^{-\frac{n}{2}} e^{ix^2/4t}\, [M_k(-\frac{\Delta}{4t})\tilde{h}]\,(\frac{x}{2t}) \tag{12.31}$$

and

$$[B_k(t)h]\,(x) = (4\pi it)^{-\frac{n}{2}} e^{ix^2/4t} \int R_k(\frac{y^2}{4t})\, e^{-i\frac{x.y}{2t}} h(y)\, dy\,. \tag{12.32}$$

It is clear that a restriction on the support of $\tilde{h}$ leads to a restriction on the support of $A_k(t)h$, since $M_k(-\frac{\Delta}{4t})$ is a differential operator. On the other hand, we have the following strong decay result for $B_k(t)h$.

LEMMA 12.8 : Given two integers ℓ and M, there exist an integer k, a number $N < \infty$ and a constant $c < \infty$ such that for $t \geq 1$ and all $h \in \mathcal{S}(\mathbb{R}^n)$:

$$t^{\ell}\|(1 + |Q|)^M B_k(t)h\|_{L^2} \leq c \sum_{|\beta|\leq N} \|(1 + |Q|)^N P^{\beta} h\|_{L^2} . \qquad (12.33)$$

Here $\beta = (\beta_1,\ldots,\beta_n)$ means a multi-index, i.e. an n-tuple of non-negative integers, $P^{\beta} = P_1^{\beta_1} \ldots P_n^{\beta_n}$ and $|\beta| = \beta_1 + \beta_2 +\ldots+ \beta_n$.

Proof :

$$\|x^{\beta}B_k(t)h\| = \|x^{\beta} (2t)^{-\frac{n}{2}} F(R_k(\tfrac{y^2}{4t})h)(\tfrac{x}{2t})\|$$

$$= \|(2tx)^{\beta} F(R_k(\tfrac{y^2}{4t})h)(x)\|$$

$$= \|(2tD_y)^{\beta} R_k(\tfrac{y^2}{4t}) h(y)\|$$

$$\leq c\ t^{|\beta|} \sum_{\alpha_1+\alpha_2=\beta} \|D_y^{\alpha_1} R_k(\tfrac{y^2}{4t}).D_y^{\alpha_2} h(y)\| , \qquad (12.34)$$

where α_1 and α_2 are multi-indices and $D_y = \mathrm{grad}_y = iP$. Now

$$D_y^{\alpha_1} R_k(\tfrac{y^2}{4t}) = \sum_{\gamma_1+\ldots+\gamma_\mu=\alpha_1} c_{\gamma} D_y^{\gamma_1}(\tfrac{y^2}{4t})\ldots D_y^{\gamma_\mu}(\tfrac{y^2}{4t}) R_k^{(\mu)}(\tfrac{y^2}{4t}) \qquad (12.35)$$

where, if $|\alpha_1| \geq 1$, each γ_s in the sum satisfies $|\gamma_s| \geq 1$. Of course $|\gamma_s| \leq 2$. If we suppose $\gamma_1 =\ldots= \gamma_\nu = 1$, $\gamma_{\nu+1} =\ldots= \gamma_\mu = 2$, the summand in (12.35) is dominated by

$$c|\frac{|y|^{\nu}}{t^{\nu}}. \frac{1}{t^{\mu-\nu}} R_k^{(\mu)}(\tfrac{y^2}{4t})| \leq c\ t^{-\mu} |y|^{\nu} |\tfrac{y^2}{4t}|^{k+1-\mu} \qquad (\text{if } \mu \leq k+1)$$

$$\leq c'\ t^{-(k+1)} |y|^{\nu+2(k+1)-\mu}.$$

Consequently

$$|D_y^{\alpha_1} R_k(\frac{y^2}{4t})| \le ct^{-(k+1)} (1 + |y|)^{|\beta|+2(k+1)} \text{ , if } |\beta| \le k+1 .$$

Combining (12.34) and (12.35) yields

$$\|x^\beta B_k(t)h\| \le ct^{|\beta|-(k+1)} \sum_{|\alpha_2|\le|\beta|} \|(1 + |y|)^{|\beta|+2(k+1)} D_y^{\alpha_2} h\| \text{ ,}$$

if $|\beta| \le k+1$, from which the lemma is an immediate consequence. ∎

Proof of Proposition 12.7 : To prove (12.25), it suffices to show that, for each $\ell,m \ge 0$:

$$\|F_{|x|\le\varepsilon(\nu+t)} e^{-iK_0 t} (K_0+i) G(\Lambda_\mu) \eta_j(P) g_\nu\| \le c_{\ell m} |\mu|^{-\ell} t^{-m} \tag{12.36}$$

if $\mu \div (\nu,j)$, $\nu \ge \nu_0$ and $t > 0$. Indeed, since $|\mu| \ge \nu^2$, (12.36) implies that for $t > 0$ and $\nu \ge \nu_0$ (take $m = 0$ in (12.36) for $0 < t \le 1$) :

$$\|F_{|x|\le\varepsilon(\nu+t)} e^{-iK_0 t} (K_0+i) g_\nu(j,out)\| \le c'_{\ell m} \nu^{-2} \sum_{\mu\div(\nu,j)} |\mu|^{-\ell+1}(1+t)^{-m} \text{ ,} \tag{12.37}$$

from which (12.25) follows since the sum over μ is finite if $\ell-1 > n$.

Let $T_\mu \equiv e^{-i\mu.P}$ be the operator of translation by the vector $\mu \in \mathbb{R}^n$ in configuration space, i.e. $(T_\mu h)(x) = h(x-\mu)$. Since T_μ is unitary and commutes with P and K_0, we have

$$\| F_{|x|\leq\varepsilon(\nu+t)} \, e^{-iK_o t} (K_o+i) \, G(\Lambda_\mu) \, \eta_j(P) \, g_\nu \|$$

$$= \| T_{-\mu} \, F_{|x|\leq\varepsilon(\nu+t)} \, T_\mu \, e^{-iK_o t} (K_o+i) \, T_{-\mu} \, G(\Lambda_\mu) \, T_\mu \, \eta_j(P) \, T_{-\mu} g_\nu \|$$

$$= \| F_{|x+\mu|\leq\varepsilon(\nu+t)} \, e^{-iK_o t} (K_o+i) \, G(\Lambda_o) \, \eta_j(P) \, T_{-\mu} g_\nu \| . \qquad (12.38)$$

Let us set $h_{\nu\mu} = (K_o+i)\, G(\Lambda_o)\, \eta_j(P)\, T_{-\mu} g_\nu$ and write $\exp(-iK_o t) = A_k(t)+B_k(t)$ in (12.38). We first show that the term arising from $A_k(t)$ is zero if ν is sufficiently large. To see this, we remark that, in the momentum variable k, the operator $G(\Lambda_o)$ is convolution by the Fourier transform of $\eta * \chi_{\Lambda_o}$, i.e. by $\tilde{\eta}(k)\, \tilde{\chi}_{\Lambda_o}(k)$. Now the support of this function lies in $|k| \leq a$, whereas the support of $[F\eta_j(P)\, T_{-\mu} g_\nu](k)$ is contained in the set $\{k \mid k.e_j \geq -a \text{ and } |k| \geq 4a\}$. It follows that

$$\mathrm{Supp}(Fh_{\nu\mu})(k) \subseteq \{k \mid k.e_j \geq -2a \ , \ |k| \geq 3a\} .$$

In view of (12.31), this implies that

$$\mathrm{Supp}(A_k(t)h_{\nu\mu})(x) \subseteq \{x \mid \tfrac{x}{2t}.e_j \geq -2a \ , \ |\tfrac{x}{2t}| \geq 3a\}.$$

In (12.38), we have to act on the vector $A_k(t)h_{\nu\mu}$ by the operator $F_{|x+\mu|\leq\varepsilon(\nu+t)}$. To verify our claim that this gives the zero vector, it suffices to show that, for $t > 0$ and $\nu \geq \nu_o$:

$$\{x \mid \tfrac{x}{2t}.e_j \geq -2a \ , \ |\tfrac{x}{2t}| \geq 3a\} \cap \{x \mid |\tfrac{x}{2t} + \tfrac{\mu}{2t}| \leq \tfrac{\varepsilon}{2} (\tfrac{\nu}{t} + 1)\} = \emptyset. \qquad (12.39)$$

Setting $z = \frac{x}{2t}$, $\theta = \frac{\mu}{2t}$, $\alpha = \frac{\nu}{t}$ and noticing that $|\theta| \geq \frac{1}{2}\alpha\nu$ (since

$|\mu| \geq \nu^2)$, (12.39) reduces to

$$\{z \mid z.e_j \geq -2a \ , \ |z| \geq 3a\} \cap \{z \mid |z+\theta| < \tfrac{\varepsilon}{2}(\alpha+1)\} = \emptyset \qquad (12.40)$$

for all $\{\alpha,\theta\} \in \mathbb{R} \times \mathbb{R}^n$ such that $\alpha > 0$, $|\theta| \geq \frac{1}{2}\alpha\nu$ and $\theta.e_j \geq \frac{99}{100}|\theta|$. Under these restrictions on α and θ , (12.40) is satisfied if $\varepsilon \leq \min(1,\frac{a}{2})$ and $\nu \geq 100$, since the second set in (12.40) is contained in the cone $C = \{z \mid (z-2ae_j).e_j \leq -\frac{9}{10}|z - 2ae_j|\}$ (with apex at the point $2a\,e_j$), whereas the first set is disjoint from C.

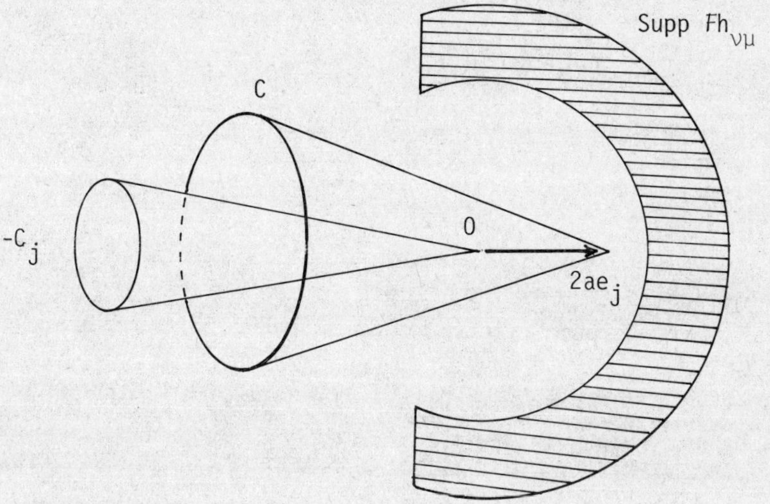

We have thus shown that, if $t > 0$ and $\nu \geq 100$:

$$\| F_{|x| \leq \varepsilon(\nu+t)} e^{-iK_0 t} (K_0+i) G(\Lambda_\mu) \eta_j(P) g_\nu \|$$

$$= \| F_{|x+\mu| \leq \varepsilon(\nu+t)} B_k(t) h_{\nu\mu} \|$$

$$\leq t^{-s} \| (I + |Q|)^{-r} F_{|x+\mu| \leq \varepsilon(\nu+t)} \| . \, t^s \, \| (I + |Q|)^r B_k(t) h_{\nu\mu} \| .$$

To estimate the last expression, we distinguish the two cases $\mu < 5t$ and $\mu \geq 5t$ and use Lemma 12.8 in each case. If $\mu < 5t$, we take $r = 0$, $s = \ell+m$ and obtain the bound

$$t^{-\ell-m} c_{\ell m} \sum_{|\beta| \leq N} \| (I + |Q|)^N P^\beta h_{\nu\mu} \|$$

$$\leq t^{-m} \left(\frac{\mu}{5}\right)^{-\ell} c_{\ell m} \sum_{|\beta| \leq N} \| (I + |Q|)^N P^\beta h_{\nu\mu} \| . \qquad (12.41)$$

If $\mu \geq 5t$, we take $r = \ell$, $s = m$ and notice that

$$|x+\mu| \leq \varepsilon(\nu+t) \quad , \quad \mu \geq 5t \quad \text{and} \quad \mu \geq \nu^2 \geq \nu_0^2 \; , \; \varepsilon \leq 1$$

implies

$$\varepsilon(\nu+t) \leq \frac{1}{2} \mu \quad \text{and} \quad |x| \geq \frac{1}{2} \mu .$$

Thus $\| (I + |Q|)^{-\ell} F_{|x+\mu| \leq \varepsilon(\nu+t)} \| \leq (1 + \frac{1}{2} \mu)^{-\ell}$, and we get a bound of the form (12.41) also in this case.

To complete the proof of (12.36), it remains to show that, for each fixed N and each $|\beta| \leq N$:

$$\sup_{\mu \geq \nu^2 \geq \nu_o^2} \|(I + |Q|)^N P^\beta h_{\nu\mu}\| \leq c_{N\beta} < \infty. \tag{12.42}$$

Now

$$\|(I+|Q|)^N P^\beta h_{\nu\mu}\| = \|(I+|Q|)^N P^\beta (K_o+i)\, G(\Lambda_o)\, \eta_j(P)\, \tilde{\phi}(K_o)\, e^{+i\mu.P} f_\nu\|$$

$$\leq \|(I+|Q|)^N P^\beta (|P|^2 + i)\, G(\Lambda_o)\, \eta_j(P)\, \tilde{\phi}(K_o)\|\, \|f\|. \tag{12.43}$$

(12.42) follows since the r.h.s. of (12.43) is independent of ν and μ and finite (by writing $P = -i\, \mathrm{grad}_x$ and commuting $P^\beta(|P|^2 + i)$ through the factor $G(\Lambda_o)$, one sees that the operator appearing in the first norm of the r.h.s. of (12.43) is a finite sum of terms of the form $(I + |Q|)^N \psi(Q) P^\alpha \eta_j(P)\tilde{\phi}(K_o)$, which is in $\mathcal{B}(\mathcal{H})$ since ψ, η_j and $\tilde{\phi}$ are in $\mathcal{S}(\mathbb{R}^n)$.)

We have thus proven (12.25). The proof of (12.26) is almost identical. The only difference arises in (12.39) which becomes

$$\{x \mid \tfrac{x}{2t}.e_j \leq 2a\ ,\ |\tfrac{x}{2t}| \geq 3a\} \cap \{x \mid |\tfrac{x}{2t} + \tfrac{\mu}{2t}| \leq \tfrac{\varepsilon}{2}\,(\tfrac{\nu}{|t|} + 1)\} = \emptyset. \tag{12.39'}$$

Setting $z = -\frac{x}{2t}$, $\theta = -\frac{\mu}{2t}$ and noticing that $t < 0$, this again reduces to (12.40). ∎

E. EXISTENCE OF THE WAVE OPERATORS.

We take an opportunity here to show how the decomposition (12.30) and the estimate (12.33) lead to the existence of the wave operators $\Omega_\pm$. We recall from Proposition 6.13 and Corollary 6.11 that Ω_+ exists provided, for h in a dense subset $\mathcal{D}$ of $L^2(\mathbb{R}^n)$,

$$\int_1^\infty \|V e^{-iK_o t} h\|\, dt < \infty, \tag{12.44}$$

and that for $g \in S(\mathbb{R}^n)$

$$\|V e^{-iK_0 t} g\| \le \|V\|_{L^2} \|e^{-iK_0 t} g\|_{L^\infty} \le c_g |t|^{-\frac{n}{2}} \|V\|_{L^2} . \qquad (12.45)$$

PROPOSITION 12.9 : If $V \in L^2_{Loc}(\mathbb{R}^n)$ and $|V(x)| \le c|x|^{-1-\varepsilon}$ for $|x| \ge R$, then $\Omega_\pm$ exist on $L^2(\mathbb{R}^n)$, $n \ge 3$.

Proof : We prove (12.44) for $h \in S(\mathbb{R}^n)$ such that $\tilde{h}(k) = 0$ for k in a neighbourhood of zero (say for $|k| \le a$). We say $h \in \mathcal{D}_a$; $\mathcal{D} = \bigcup_{a>0} \mathcal{D}_a$ is dense in $L^2(\mathbb{R}^n)$. This will prove that Ω_+ exists, and the existence of Ω_- follows by a similar argument. Indeed, let us write

$$V = VF_{|x|\le R} + VF_{|x|\ge R} = V_1 + V_2 ,$$

$$\|V e^{-iK_0 t} h\| \le \|V_2 A_k(t)h\| + \|V_2 B_k(t)h\| + \|V_1 e^{-iK_0 t} h\|. \qquad (12.46)$$

Since $A_k(t)\, h(x) = 0$ for $|x| \le 2at$, $h \in \mathcal{D}_a$, we have

$$\|V_2 A_k(t)h\| \le c(2at)^{-1-\varepsilon} \|A_k(t)h\| ,$$

while

$$\|A_k(t)h\| \le \|e^{-iK_0 t} h\| + \|B_k(t)h\| \le c .$$

Thus

$$\|V_2 A_k(t)h\| \le c\,(at)^{-1-\varepsilon} . \qquad (12.47)$$

The estimate (12.33) yields

$$\|V_2\, B_k(t)h\| \le c\, \|B_k(t)h\| \le c_\ell\, t^{-\ell}\,. \tag{12.48}$$

Finally the estimate (12.45) applies to $V_1 \in L^2$, so

$$\|V_1\, e^{-iK_o t} h\| \le c_h\, t^{-\frac{n}{2}}\, \|V_1\|_{L^2}\,. \tag{12.49}$$

Now (12.47) - (12.49) applied to (12.46) yield

$$\|V\, e^{-iK_o t} h\| \le c_h'\, t^{-1-\varepsilon}\quad,\quad h \in \mathcal{D}_a\,,$$

so the proof is complete. ■

F. COMPLETENESS OF THE WAVE OPERATORS.

We begin by showing that for ν large, $\Omega_\pm$ are approximatively the identity on $g_\nu(j\,,\text{out/in})$.

LEMMA 12.10 : We have

$$\lim_{\nu\to\infty} \|(\Omega_+ - I)\, g_\nu(j\,,\text{out})\| = 0 \tag{12.50}$$

$$\lim_{\nu\to\infty} \|(\Omega_- - I)\, g_\nu(j\,,\text{in})\| = 0\,. \tag{12.51}$$

Proof :

$$\begin{aligned}\|(\Omega_+ - I)\, g_\nu(j\,,\text{out})\| &\le \int_0^\infty \|V\, e^{-iK_o t}\, g_\nu(j\,,\text{out})\|\, dt \\ &\le \|V(K_o+i)^{-1}\| \int_0^\infty \|F_{|x|\le\varepsilon(\nu+t)}\, e^{-iK_o t}\,(K_o+i)\, g_\nu(j\,,\text{out})\|\, dt \\ &\quad + \|(K_o+i)\, g_\nu(j\,,\text{out})\| \int_0^\infty \|V(K_o+i)^{-1}\, F_{|x|\ge\varepsilon(\nu+t)}\|\, dt.\end{aligned} \tag{12.52}$$

The first term in the sum vanishes as $\nu \to \infty$ by Proposition 12.7. For the second term, we notice that $\tilde{g}_\nu(j, \text{out})$ has support in $\{k \mid |k| \le b + a\}$, since it is a sum of terms each of which has its support in this set (by the argument following (12.38).) Hence

$$\|(K_0+i)\, g_\nu(j, \text{out})\| \le [(b+a)^2 + 1]\, \|g_\nu(j, \text{out})\| .$$

Now

$$\|g_\nu(j, \text{out})\| \le \|\sum_{\mu:(\nu,j)} G(\Lambda_\mu)\|\, \|\eta_j\|_\infty\, \|\tilde{\phi}\|_\infty\, \|f\|$$

and

$$\|\sum_{\mu\div(\nu,j)} G(\Lambda_\mu)\| = \sup_{x\in \mathbb{R}^n} \left|\sum_{\mu\div(\nu,j)} \int \eta(x-y)\, \chi_{\Lambda_\mu}(y)\, dy\right| \le \|\eta\|_1 .$$

Hence $\|(K_0+i)\, g_\nu(j, \text{out})\|$ is bounded by a constant which is independent of ν. The convergence to zero of the second term in (12.52) now follows from the short range hypothesis (12.2) and the Lebesgue dominated convergence theorem, since

$$\int_0^\infty \|V(K_0+i)^{-1}\, F_{|x|\ge\varepsilon(\nu+t)}\|\, dt \le \int_0^\infty \|V(K_0+i)^{-1}\, F_{|x|\ge\varepsilon t}\|\, dt$$

$$= \varepsilon^{-1} \int_0^\infty \|V(K_0+i)^{-1}\, F_{|x|\ge\tau}\|\, d\tau .$$

This proves (12.50). (12.51) is proved similarly. ∎

We need one more technical lemma to the effect that $g_\nu(j, \text{out/in})$ and f_ν are almost orthogonal for large ν.

LEMMA 12.11 :

$$\lim_{\nu\to\infty} |(g_\nu(j, \text{in}),\, f_\nu)| = 0 . \qquad (12.53)$$

Proof : Since $e^{-iK_0t}\,\Omega_-^* = \Omega_-^*\,e^{iHt}$, we have

$$|(g_\nu(j\,,\mathrm{in}),f_\nu)| \le \|(1-\Omega_-)\,g_\nu(j\,,\mathrm{in})\|\,\|f\|$$

$$+\,|(e^{iK_0\tau_\nu}(K_0+i)\,g_\nu(j\,,\mathrm{in})\,,\,\Omega_-^*(H+i)^{-1}f)|.$$

The first summand vanishes as $\nu \to \infty$ by Lemma 12.10 and the second one is bounded by

$$\|F_{|x|\le\varepsilon(\nu+\tau_\nu)}\,e^{iK_0\tau_\nu}(K_0+i)\,g_\nu(j\,,\mathrm{in})\|\,\|f\|$$

$$+\,\|F_{|x|>\varepsilon(\nu+\tau_\nu)}\,\Omega_-^*(H+i)^{-1}f\|\,\|(K_0+i)\,g_\nu(j\,,\mathrm{in})\|.$$

The first term vanishes as $\nu \to \infty$ by the analogue of (12.37) for $g_\nu(j\,,\mathrm{in})$, and the last term vanishes since $\|(K_0+i)\,g_\nu(j\,,\mathrm{in})\| \le c$ for all ν and since $\text{s-lim}\,F_{|x|\ge R} = 0$ as $R \to \infty$. ∎

PROPOSITION 12.12 : On the short range hypothesis (12.2) on the potential V, the wave operators $\Omega_\pm$ are complete in $L^2(\mathbb{R}^n)$, $n \ge 3$, in the sense that range $\Omega_\pm = \mathcal{H}_c(H)$.

Proof : Suppose $f \in \mathcal{H}_c(H)$ is orthogonal to the range of Ω_+. It follows that $\phi(H)f$ shares this property, for any $\phi \in C_0^\infty(\mathbb{R})$ so without loss of generality we can assume f has energy support in an interval $\Delta = [16a^2, b^2]$ as in section C. Let $f_\nu = e^{-i\tau_\nu H}f$ as in (12.10) ; also f_ν is orthogonal to $R(\Omega_+)$. We want to show that f = 0. Indeed,

$$\|f\|^2 = \|f_\nu\|^2$$

$$= \lim_{\nu\to\infty}\,(f_\nu - \sum_j g_\nu(j\,,\mathrm{in}),f_\nu) \tag{12.54}$$

by Lemma 12.11, and

$$(f_\nu - \sum_j g_\nu(j\,,\text{in})\ ,\ f_\nu)$$

$$= (\Omega_+ \sum_j g_\nu(j\,,\text{out})\ ,\ f_\nu)$$

$$+ ((I - \Omega_+) \sum_j g_\nu(j\,,\text{out})\ ,f_\nu)$$

$$+ (f_\nu - \sum_j g_\nu(j\,,\text{in}) - \sum_j g_\nu(j\,,\text{out})\ ,f_\nu)\ . \qquad (12.55)$$

Since $f_\nu \perp \mathcal{R}(\Omega_+)$, the first term on the r.h.s. of (12.55) vanishes for all ν. Lemma 12.10 implies that the second term vanishes as $\nu \to \infty$, whereas the third term converges to zero by (12.24). This shows that f = 0, so range $\Omega_+ = \mathcal{H}_c(H)$. Similarly one obtains that range $\Omega_- = \mathcal{H}_c(H)$.∎

Remark : The short range hypothesis (12.2) is satisfied for instance if $V \in L^{\frac{n}{2}}_{Loc}(\mathbb{R}^n)$ and $|V(x)| \le (1 + |x|)^{-1-\varepsilon}$ for $|x| \ge R$,[35].

APPENDIX : MODIFIED WAVE OPERATORS FOR LONG RANGE POTENTIALS.

If the potential $V(x)$ fails to decrease like $|x|^{-1-\varepsilon}$ as $|x| \to \infty$, as in the Coulomb case $V(x) = c|x|^{-1}$, the wave operators $\Omega_\pm$ will in general fail to exist. However, modified wave operators

$$\hat{\Omega}_\pm = \underset{t\to\pm\infty}{\text{s-lim}}\, e^{iHt} e^{-iK_0 t - iW(t)} \tag{12.56}$$

may exist, for a certain correction $W(t)$ which we shall describe shortly. It turns out that the technical Lemma 12.8, which amongst other things led to a quick proof of the existence of $\Omega_\pm$ in the short range case (Theorem 12.9), is also a convenient tool for proving the existence of the modified wave operators $\hat{\Omega}_\pm$, under some hypotheses and we take the opportunity here to give an exposition of this. We will not discuss the question of the completeness of the modified wave operators.

We shall make the following hypotheses on the potential V : there is a number $\varepsilon \in (0,\frac{1}{2}]$ such that

$$|V(x)| \le c|x|^{-\frac{1}{2}-\varepsilon} \qquad \text{for} \quad |x| \ge R \,, \tag{12.57}$$

$$|D^\beta V(x)| \le c_\beta |x|^{-1-\frac{1}{2}|\beta|-\varepsilon|\beta|} \qquad \text{for} \quad |x| \ge R \quad \text{and} \quad |\beta| \ge 1 \,, \tag{12.58}$$

and we make the local hypothesis :

$$V \in L^p_{Loc}(\mathbb{R}^n) \quad , \quad p \ge 2 \quad , \quad p > \frac{n}{2} \quad . \tag{12.59}$$

Note that hypothesis (12.58) implies (12.57) ; we state (12.57) only for emphasis.

For $W(t)$ we take

$$W(t) = \int_0^t \hat{V}(2\sigma P)\, d\sigma, \tag{12.60}$$

where $\hat{V}(x) = V(x)$ for $|x| \geq R$ and $\hat{V}(x)$ smooth for $|x| \leq R$. With a more subtle choice of $W(t)$, one can relax the conditions (12.57) and (12.58) considerably, as we shall further mention at the end of this appendix.

As in the existence proof of $\Omega_\pm$ for short range potentials, we begin by differentiating

$$P(t) = e^{iHt}\, e^{-iK_0 t - iW(t)}$$

to get

$$P'(t) = i\, e^{iHt}\, [V - \hat{V}(2tP)]\, e^{-iK_0 t - iW(t)}.$$

By Lemmas 6.8 and 6.9 we have :

<u>LEMMA 12.13</u> : $\hat{\Omega}_+$ exists provided, for f in a dense subset $\mathcal{D}$ of $L^2(\mathbb{R}^n)$,

$$\int_1^\infty \| [V - \hat{V}(2tP)]\, e^{-iK_0 t}\, e^{-iW(t)} f \|\, dt < \infty. \tag{12.61}$$

So our task is to establish (12.61), which we shall now do. We set

$$f_t = e^{-iW(t)} f \tag{12.62}$$

and define $\mathcal{D}_{a,b}$ $(0 < a < b < \infty)$ to be the set of all functions $f : \mathbb{R}^n \to \mathbb{C}$ such that $\tilde{f}$ is C^∞ and supp $\tilde{f} \subseteq \{k \mid a \leq |k| \leq b\}$. It suffices to prove (12.61) for each f belonging to some $\mathcal{D}_{a,b}$, since $\mathcal{D} = \bigcup_{0<a<b<\infty} \mathcal{D}_{a,b}$ is dense in $L^2(\mathbb{R}^n)$.

LEMMA 12.14 : Let $0 < a < b < \infty$, $f \in \mathcal{D}_{a,b}$ and $s > 0$ an integer. Then there is a constant c such that, for $t \geq \max\{1, (2a)^{-1}R\}$:

$$\| |Q|^{2s} f_t \| \leq c\, t^{s-2\varepsilon s} , \tag{12.63}$$

$$\| |Q|^{2s} \hat{V}(2tP) f_t \| \leq c\, t^{s-2\varepsilon s} , \tag{12.64}$$

$$\| [\hat{V}(2tP)\, |Q|^{2s} - |Q|^{2s}\, \hat{V}(2tP)] f_t \| \leq c\, t^{s-2\varepsilon s-1} . \tag{12.65}$$

Proof : We have

$$\| |Q|^{2s} \hat{V}(2tP) f_t \| = \| \Delta_k^s\, \hat{V}(2tk)\, e^{-iW(t,k)}\, \tilde{f}(k) \| . \tag{12.66}$$

Now $\Delta_k^s\, \hat{V}(2tk)\, e^{-iW(t,k)}\, \tilde{f}(k)$ is a finite linear combinations of terms of the form

$$e^{-iW(t,k)}\, [D_k^\alpha\, \hat{V}(2tk)]\, [D_k^\beta\, \tilde{f}(k)]\, [D_k^{\gamma_1}\, W(t,k)]^{n_1} \ldots [D_k^{\gamma_m}\, W(t,k)]^{n_m} \tag{12.67}$$

with

$$|\alpha| + |\beta| + \sum n_i |\gamma_i| = 2s . \tag{12.68}$$

From (12.58) we get for $|k| \geq a$ and $t \geq R/2a$:

$$\begin{aligned} |D_k^\alpha\, \hat{V}(2tk)| &\leq (2t)^{|\alpha|}\, c_\alpha\, (2tk)^{-1-\frac{1}{2}|\alpha|-\varepsilon|\alpha|} \\ &\leq c(\alpha,a)\, t^{-1-\frac{1}{2}|\alpha|-\varepsilon|\alpha|} , \quad (\alpha \neq 0) \end{aligned} \tag{12.69}$$

$$|D_k^\beta \tilde{f}(k)| \le c(\beta,f) \tag{12.70}$$

and

$$\begin{aligned}|D_k^\gamma W(t,k)| &\le \int_0^t d\tau \; |D_k^\gamma \hat{V}(2\tau k)| = \int_0^{R/2a} \ldots + \int_{R/2a}^t \ldots \\ &\le c_1(\gamma,a) + c(\gamma,a)\int_{R/2a}^t \tau^{-1+\frac{1}{2}|\gamma|-\varepsilon|\gamma|} d\tau \\ &\le c_2(\gamma,a)\; t^{\frac{1}{2}|\gamma|-\varepsilon|\gamma|} \; .\end{aligned} \tag{12.71}$$

Hence, if $\alpha \neq 0$, the absolute value of the function (12.67), for $t \ge \max\{1,(2a)^{-1}R\}$, is bounded by

$$\begin{aligned}&c_3(\alpha,\beta,\gamma_1,\ldots,\gamma_m,f)\; t^{-1+\frac{1}{2}|\alpha|-\varepsilon|\alpha|+\frac{1}{2}\sum n_i|\gamma_i|-\varepsilon\sum n_i|\gamma_i|} \\ &\quad = c_3\; t^{-1+s-\frac{1}{2}|\beta|-\varepsilon(2s-|\beta|)} \le c_3\; t^{s-2\varepsilon s-1} \\ &\quad \le c_3\; t^{s-2\varepsilon s} \; ,\end{aligned} \tag{12.72}$$

where we have used (12.68) and the hypothesis that $\varepsilon \le \frac{1}{2}$.

Similarly, if $\alpha = 0$, one finds from (12.57), (12.70) and (12.71) that the corresponding term is bounded, as in (12.72), by $c\; t^{s-2\varepsilon s}$.

Inserting the estimate (12.72) into (12.66) and using the fact that $\tilde{f}$ has support in the *bounded* set $\{k \mid |k| \le b\}$, one arrives at (12.64). (12.65) is obtained using the bound given by the first inequality in (12.72) ; this is permissible since, due to the commutator, the function appearing in the norm in (12.65) is a linear combination of terms of the form (12.67) with $\alpha \neq 0$.

The proof of (12.63) is quite similar. It suffices to replace in (12.67) the factor $D_k^\alpha \hat{V}(2tk)$ by 1 and to put $|\alpha| = 0$ in (12.68). ∎

In the next lemma we use the operator $B_r(t)$ defined in (12.32).

LEMMA 12.15 : Let $0 < a < b < \infty$, $f \in \mathcal{D}_{a,b}$, $\ell > 0$ and $r > (2\varepsilon)^{-1}\ell$. Then there is a constant c such that, for $t \geq \max\{1, (2a)^{-1}R\}$:

$$\|B_r(t)\ \hat{V}(2tP)f_t\| \leq c\ t^{-\ell}\ , \tag{12.73}$$

and

$$\|V\ B_r(t)f_t\| \leq c\ t^{-\ell}. \tag{12.74}$$

Proof : Using (12.34) with $\beta = 0$ and (12.29) with $\ell = 0$, we find that for all $h \in \mathcal{S}(\mathbb{R}^n)$

$$\|B_r(t)\ h\| \leq \text{const.}\ t^{-(r+1)}\ \||Q|^{2(r+1)}\ h\|\ . \tag{12.75}$$

(12.73) follows immediately upon taking $h = \hat{V}(2tP)f_t$ and using (12.64).

Similarly, if we set $h = f_t$ in (12.75) and use (12.63), we obtain

$$\|B_r(t)f_t\| \leq c(r,f)\ t^{-\ell} \quad \text{for} \quad r > \frac{\ell}{2\varepsilon}\ . \tag{12.76}$$

It follows that, for $r \geq (2\varepsilon)^{-1}\ell$,

$$\begin{aligned}\|V\ B_r(t)f_t\| &\leq \|\hat{V}\ B_r(t)f_t\| + \|(V-\hat{V})\ B_r(t)f_t\| \\ &\leq c\|\hat{V}\|_\infty\ t^{-\ell} + \|(V-\hat{V})(K_0+1)^{-1}\|\ [\|K_0B_r(t)f_t\| + c\ t^{-\ell}]\ .\end{aligned}$$

Since $V(.) - \hat{V}(.) \in L^p(\mathbb{R}^n)$ with $p \geq 2$, $p > \frac{n}{2}$, $(V-\hat{V})(K_0+1)^{-1}$ is a bounded operator by Proposition 2.8. To prove (12.74), it therefore suffices to show that

$$\|K_0 \, B_r(t) f_t\| \leq c \, t^{-\ell} \quad \text{if} \quad r \geq (2\varepsilon)^{-1}\ell. \tag{12.77}$$

Now, using the definition (12.32) of $B_r(t)$, the fact that $(K_0 g)(x) = -\Delta g(x)$ and Leibniz' rule for differentiating a product, one finds the following identity :

$$K_0 \, B_r(t)h = \left(\frac{|Q|^2}{4t^2} - \frac{in}{2t}\right) B_r(t)h - \sum_{j=1}^{n} \frac{Q_j}{2t^2} B_k(t) \, Q_j h$$

$$+ \frac{1}{4t^2} B_r(t) \, |Q|^2 h \, . \tag{12.78}$$

Upon using (12.34) with $|\beta| = 0,1,2$ and (12.29) with $\ell = 0,1,2$, we get from (12.78) the estimate

$$\|K_0 \, B_r(t)h\| \leq c \, t^{-(r+1)} \left[\sum_{|\alpha|=2} \| \, |Q|^{2r+2} \, P^\alpha h\| \right.$$

$$+ \sum_{|\alpha|=1} \| \, |Q|^{2r+1} \, P^\alpha h\| + \left. \| \, |Q|^{2r} h\| \right]$$

$$+ c \, t^{-(r+2)} \left[\| \, |Q|^{2r+1} h\| + \| \, |Q|^{2r+2} h\| + \sum_{j=1}^{n} \| \, |Q|^{2r+2} P_j Q_j h\| \right]$$

$$+ c \, t^{-(r+3)} \, \| \, |Q|^{2r+4} h\|. \tag{12.79}$$

We now set $h = f_t$ and notice that $P_j Q_j h = Q_j P_j h - ih$ and $P^\alpha f_t = (P^\alpha f)_t$, where $(FP^\alpha f)(k) = k^\alpha \, \tilde{f}(k)$. Each term in (12.79) can then be estimated by (12.63), and one thus obtains the bound (12.77). ∎

PROPOSITION 12.16 : Under the hypotheses (12.57) - (12.59), on the "moderately long range" potential V(x), the modified wave operators $\hat{\Omega}_{\pm}$ defined by (12.56) exist.

Proof : (for $\hat{\Omega}_{+}$) : Let $0 < a < b < \infty$ and $f \in \mathcal{D}_{a,b}$. By Lemma (12.13), it suffices to prove that, for some $\delta > 0$ and all $t \geq t_o$:

$$\|[V - \hat{V}(2tP)]\, e^{-iK_o t} f_t\| \leq c\, t^{-1-\delta}. \tag{12.80}$$

For this we use the decomposition (12.30) of $e^{-iK_o t}$ to write

$$[V - \hat{V}(2tP)]\, e^{-iK_o t} f_t = V\, A_r(t) f_t + V\, B_r(t) f_t$$

$$- A_r(t)\, \hat{V}(2tP) f_t - B_r(t)\, \hat{V}(2tP) f_t . \tag{12.81}$$

If $r > \varepsilon^{-1}$, the norms of the two terms involving $B_r(t)$ satisfy (12.80), by virtue of the Lemma 12.15. Also, $V\, A_r(t) f_t = \hat{V}\, A_r(t) f_t$ if $t > (2a)^{-1} R$, which follows easily from the expression (12.31) for $A_r(t)$, the definition of $\hat{V}$ and the fact that $\tilde{f}(k) = 0$ for $|k| \leq a$. Therefore it suffices to prove that, for t large and some $r > \varepsilon^{-1}$,

$$\|\hat{V}\, A_r(t) f_t - A_r(t)\, \hat{V}(2tP) f_t\| \leq c\, t^{-1-\delta}. \tag{12.82}$$

Let us define $\hat{V}_t(\xi) = \hat{V}(2t\xi)$. Then, from (12.31)

$$[\hat{V}\, A_r(t) f_t - A_r(t)\, \hat{V}(2tP) f_t]\,(x)$$

$$= (4\pi i t)^{-\frac{n}{2}}\, e^{i\frac{x^2}{4t}}\, \{[\hat{V}_t(.), M_r(-\tfrac{\Delta}{4t})]\, \tilde{f}_t(.)\}\, (\tfrac{x}{2t}) ,$$

where $[A,B] \equiv AB - BA$. Thus, by the change of variables $x \mapsto k = \frac{x}{2t}$:

$$\|\hat{V}\, A_r(t) f_t - A_r(t)\, \hat{V}(2tP) f_t\|$$

$$= (2\pi)^{-\frac{n}{2}}\ \|[\hat{V}_t(k),\, M_r(-\tfrac{\Delta_k}{4t})]\tilde{f}_t(k)\|_{L^2(\mathbb{R}^n)} .$$

Since $M_r(u)$ is a polynomial in u, it suffices to show that

$$\|[\hat{V}_t(k)\ ,\ \Delta_k^s]\ \tilde{\dot{f}}_t(k)\| \leq c\ t^{s-1-\delta} \tag{12.83}$$

for each positive integer s and for t sufficiently large. Since $|Q|^2 = -\Delta_k$, (12.83) with $\delta = 2\varepsilon$ follows directly from (12.65). ∎

Of course, the modified wave operators $\hat{\Omega}_\pm$ defined by (12.56) are easily shown to intertwine e^{iHt} and $e^{iK_0 t}$:

$$e^{iHt}\ \hat{\Omega}_\pm = \hat{\Omega}_\pm\ e^{iK_0 t} . \tag{12.84}$$

Work on such modified wave operators goes back to Dollard, [18], see also Alsholm and Kato [4], Alsholm [3] and Buslaev and Matveev [14]. The latter two allow V with $|D^\alpha V(x)| \leq c(1+|x|)^{-|\alpha|-\varepsilon}$; arranging this requires modification of the formula (12.60). Hörmander in [23] obtains still more general results replacing (12.56) by

$$\hat{\Omega}_\pm = \underset{t\to\pm\infty}{\text{s-lim}}\ e^{itH}\ e^{-iW_1(t,P)}$$

where, rather than having $W_1(t,P) = tK_0 + W(t)$ defined by (12.60), $W_1(t,K)$ is obtained as an exact solution to

$$V(\nabla_k W_1) + |k|^2 = \frac{\partial}{\partial t} W_1 ,$$

at least for $(|t|,k) \in [T_K,\infty) \times K$ for every compact $K \subset \mathbb{R}^n \setminus \{0\}$. For details, see Hörmander [23]. The proof given in these notes in based on [12].

13 Scattering theory

In this chapter we give some physical applications of Schrödinger operators. We first explain their interpretation in Quantum Mechanics, then describe Scattering Theory in terms of the wave operators and finally discuss the most important physical quantity, namely the scattering cross section.

A. QUANTUM MECHANICS.

In Classical Mechanics, a *particle* is considered to be a point-like object having a mass m. The *state* of a particle is characterized by giving its *position* $x \in \mathbb{R}^n$ and its *velocity* $v \in \mathbb{R}^n$.

If the real variable t is interpreted as the *time* the function $t \mapsto x(t)$, where $x(t)$ is the position at the time t, corresponds to the trajectory of the particle in physical space $\mathbb{R}^n$. The velocity is $v(t) = \frac{dx(t)}{dt}$. The quantity $p = mv$ is called the *momentum* of the particle.

The *equation of motion* is a second-order differential equation (Newton's equation)

$$m \frac{d^2x}{dt^2} = F(x) \quad ,$$

where F is the external force acting on the particle.

Example : F is derived from a potential V(x) :

$$F(x) = - \operatorname{grad} V(x) \quad .$$

V is called the *potential energy*, $(2m)^{-1} p^2$ the *kinetic energy* and $(2m)^{-1} p^2 + V$ the *total energy*. $(2m)^{-1} p^2 + V(x)$ is constant on each trajectory.

In Quantum Mechanics the above quantities become operators. The particle does not have a definite position but only a position probability density.

Postulate I :

(a) The states of the particles are the so-called wave functions $f(x)$, where $f \in L^2(\mathbb{R}^n)$, i.e. the set of states is $\mathcal{H} = L^2(\mathbb{R}^n)$.(*)

(b) $\frac{1}{\|f\|^2} \int_\Delta |f(x)|^2 \, dx \equiv P(f, \Delta)$ is the probability that the particle in the state f is localized in the region $\Delta \subseteq \mathbb{R}^n$.

Remark : If Δ is a single point, then $P(f, \Delta) = 0$, i.e. a quantum-mechanical particle is never localized at a point, as already stated above.

The multiplication operator Q_j by x_j in $L^2(\mathbb{R}^n)$ is called the j-th component of the *position operator*. Define the projection F_Δ by ($\Delta \subseteq \mathbb{R}^n$):

$$(F_\Delta f)(x) = \begin{cases} f(x) & x \in \Delta \\ 0 & x \notin \Delta . \end{cases}$$

The range of F_Δ is called the set of states which are localized in Δ with probability 1 : If

$$\begin{cases} F_\Delta f = f : P(f, \Delta) = 1 & , \\ F_\Delta f = 0 : P(f, \Delta) = 0 & . \end{cases}$$

In general

$$P(f, \Delta) = \frac{\|F_\Delta f\|^2}{\|f\|^2} ,$$

which is the probability of presence in Δ of a particle in the state f. In principle this probability can be measured experimentally and one calls F_Δ

Footnote

(*) This definition of a state is not quite correct but sufficient for our purposes. The interested reader may consult a text on basic Quantum Mechanics [J].

an *observable* (more precisely a "yes-no observable" since $\sigma(F_\Delta) = \{0,1\}$).
General definition of an observable : An observable is given by a self-adjoint operator A, or equivalently by a spectral family $\{E_\lambda\}$.

$\frac{\|E_\Delta f\|^2}{\|f\|^2}$ is the probability that in a state f the value of A is in Δ.

Postulate II : The momentum observable is given by the following n operators P_j $(j = 1,\ldots,n)$:

$$(\widetilde{P_j f})(k) = \hbar\, k_j\, \tilde{f}(k) ,$$

where $\hbar$ is Planck's constant (an experimentally determined number). Formally

$$(P_j f)(x) = - i \hbar \frac{\partial f(x)}{\partial x_j} \quad , \text{ whence}$$

$$P_j Q_\ell f - Q_\ell P_j f = - i \hbar\, \delta_{\ell j} f \quad . \tag{13.1}$$

(13.1) are called the *canonical commutation relations*.

We define the projection G_Δ by

$$(\widetilde{G_\Delta f})(k) = \begin{cases} \tilde{f}(k) & \hbar k \in \Delta \\ 0 & \hbar k \notin \Delta . \end{cases}$$

G_Δ is the observable "Momentum in Δ". In analogy with the classical case, one defines the *kinetic energy operator* by

$$\frac{1}{2m} \sum_{j=1}^{n} P_j^2 = \frac{P^2}{2m} = \frac{\hbar^2}{2m} K_o \quad .$$

From now on we set $2m = 1$ and $\hbar = 1$.

The potential energy operator is the multiplication operator by $V(x)$.
The total energy operator (called *Hamiltonian operator* in Physics) is

$$H = K_o + V \quad .$$

H must be a self-adjoint operator in order to define an observable.

Postulate III : The equation of motion for a state $f \in D(H)$ is the *Schrödinger equation*

$$i \frac{df(t)}{dt} \equiv i \operatorname{s-lim}_{\tau \to 0} \frac{f(t+\tau) - f(\tau)}{\tau} = H f(t) \quad , \quad f(0) = f \quad ,$$

where $t \mapsto f(t)$ is a trajectory of a state in Hilbert space.
In integrated form : $f(t) = e^{-iHt} f(0)$.

B. SCATTERING THEORY.

PROPOSITION 13.1 : Assume V is K_o-bounded with K_o-bound less than 1. Let $H = K_o + V$ and $f \in \mathcal{H}_{ac}(H)$. Let S_R be the ball $S_R = \{x \mid |x| \le R\}$. Then

$$\lim_{t \to \pm\infty} \| F_{S_R} \; e^{-iHt} \; f \|^2 = 0 \; .$$

In other words : A particle in a state $f \in \mathcal{H}_{ac}(H)$ disappears from each finite region as $t \to \pm\infty$.

Proof :

i) Let $f \in \mathcal{H}_{ac}(H)$, $g \in \mathcal{H}$. Since $E_\lambda E_{ac} = E_{ac} E_\lambda$:

$$(g \,, e^{-iHt} f) = \int e^{-i\lambda t} \, d(g \,, E_\lambda f)$$

$$= \int e^{-i\lambda t} \frac{d}{d\lambda}(E_{ac} g \,, E_\lambda f) \, d\lambda \; .$$

Since $E_{ac}\, g, f \in \mathcal{H}_{ac}(H)$, the Radon-Nikodym derivative $\frac{d}{d\lambda}(E_{ac} g \,, E_\lambda f)$ exists and is in $L^1(\mathbf{R})$. By the Riemann-Lebesgue lemma :

$$\lim_{t \to \pm\infty} (g \,, e^{-iHt} f) = 0 \quad ,$$

i.e. $\operatorname{w-lim}_{t \to \pm\infty} e^{-iHt} f = 0$.

ii) $$F_{S_R}(H+i)^{-1} = F_{S_R}(K_o+i)^{-1} - F_{S_R}(K_o+i)^{-1} V(H+i)^{-1}$$

by the second resolvent equation (4.8).
Now $F_{S_R}(K_o+i)^{-1}$ is compact by Proposition 2.8, and $V(H+i)^{-1} \in \mathcal{B}(\mathcal{H})$ by

Lemma 4.13(a). Hence $F_{S_R}(H+i)^{-1}$ is compact.

iii) Let $f \in \mathcal{H}_{ac}(H)$, $\varepsilon > 0$. Choose $g \in \mathcal{H}_{ac}(H) \cap D(H)$ with $\|g-f\|^2 < \frac{\varepsilon}{4}$. Then

$$\|F_{S_R} e^{-iHt} f\|^2 \leq 2 \|F_{S_R} e^{-iHt} g\|^2 + 2 \|F_{S_R} e^{-iHt}(f-g)\|^2$$

$$\leq 2 \|F_{S_R}(H+i)^{-1} e^{-iHt}(H+i)g\|^2 + \frac{\varepsilon}{2} \quad .$$

Now $(H+i)g \in \mathcal{H}_{ac}(H)$, hence $e^{-iHt}(H+i)g$ tends weakly to zero by (i). Since $F_{S_R}(H+i)^{-1} \in \mathcal{B}_\infty$, we obtain by using Lemma 4.14 that

$$\underset{|t|\to\infty}{\text{s-lim}}\, F_{S_R}(H+i)^{-1} e^{-iHt}(H+i)g = 0 \quad .$$

Hence there is $T < \infty$ such that $\|F_{S_R} e^{-iHt} f\|^2 < \varepsilon$, $\forall |t| > T$. ∎

The states in $\mathcal{H}_{ac}(H)$ are called *scattering states* of H. A scattering phenomenon is characterized as follows :
As $t \to -\infty$, a particle is far away from the center of force (given by the potential which is essentially different from zero only in a finite region). The particle moves into this "scattering region", gets deflected or scattered, and as $t \to +\infty$, it moves again far away from this region.

Since at large $|t|$, the particle is far away from the "scattering region", it should not feel any force due to the potential, i.e. it should behave like a "free" particle. By definition a particle is *free* if its time evolution is given by $\exp(-iK_0 t)$ (i.e. $V = 0$), i.e. $f(t) = \exp(-iK_0 t)f$.

A *scattering system* is such that at large positive and negative times the scattering states of H behave like free states in some appropriate sense. This is a restriction on the potential V, called the *asymptotic condition*. There are various possibilities of formulating mathematically this condition. The most usual one is in terms of convergence of states :
If f is a scattering state of H, there should exist two states $f_\pm \in \mathcal{H}$, evolving freely and approximating $\exp(-iHt)f$ as $t \to +\infty$ (resp. as $t \to -\infty$), i.e. such that

$$\lim_{t \to \pm\infty} \| e^{-iHt} f - e^{-iK_0 t} f_\pm \| = 0 \quad . \tag{13.2}$$

Now $\| e^{-iHt} f - e^{-iK_0 t} f_\pm \| = \| f - e^{iHt} e^{-iK_0 t} f_\pm \|$, so that $f = \Omega_\pm f_\pm$.

f_- is the *initial state* (at $t = 0$), f_+ the *final state* (at $t = 0$). In an experiment one prepares the initial state at large negative t, i.e. $\exp(-iK_0 t) f_-$ with $t << -1$. One measures the final state $\exp(-iK_0 t) f_+ \approx \exp(-iHt) \Omega_- f_-$ at $t >> +1$. From the measurement one wants to deduce e.g. properties of the potential (see further down).

The correspondence $f_- \mapsto f_+$ is a linear map called the *scattering operator* S. One sees that

$$S = \Omega_+^{-1} \Omega_- \quad ,$$

provided that Ω_+ is invertible on the range of Ω_-. This is the case if range $\Omega_- \subseteq$ range Ω_+, as can be seen from the following lemma the proof of which we leave as an exercise :

LEMMA 13.2 : Let Ω be an isometric operator. Then $\Omega^{-1} f$ exists if $f \in$ range Ω, and for such f

$$\Omega^{-1} f = \Omega^* f \quad .$$

Hence we have

PROPOSITION 13.3 : Assume that $\Omega_\pm$ exist and are complete. Then the pair (H , K_0) forms a scattering system. The scattering operator is given by

$$S = \Omega_+^* \Omega_- \tag{13.3}$$

and is a unitary operator commuting with $\exp(-iK_0 t)$.

Proof : Since $\Omega_\pm$ map $\mathcal{H}$ onto $\mathcal{H}_{ac}(H)$, the asymptotic condition (13.2) is verified : if $f \in \mathcal{H}_{ac}(H)$, then we take $f_\pm = \Omega_\pm^* f$.

Since range Ω_- = range Ω_+ (= $\mathcal{H}_{ac}(H)$), S is given by (13.3). Also

$$S^*S = \Omega_-^* \Omega_+ \Omega_+^* \Omega_- = \Omega_-^* E_{ac}(H) \Omega_- = \Omega_-^* \Omega_- = I \quad ,$$

$$SS^* = \Omega_+^* \Omega_- \Omega_-^* \Omega_+ = \Omega_+^* E_{ac}(H) \Omega_+ = \Omega_+^* \Omega_+ = I \quad ,$$

i.e. S is unitary. Using Proposition 6.2 and its corollary, we get

$$S e^{-iK_o t} = \Omega_+^* \Omega_- e^{-iK_o t} = \Omega_+^* e^{-iHt} \Omega_-$$

$$= e^{-iK_o t} \Omega_+^* \Omega_- = e^{-iK_o t} S \quad . \blacksquare$$

<u>COROLLARY 13.4</u> : Let $H = K_o + V$ with V satisfying (11.10). Then all conclusions of Proposition 13.3 are true.

We next derive a stationary expression for S. We set $R_z = (H - z)^{-1}$, $R_z^o = (K_o - z)^{-1}$.

<u>PROPOSITION 13.5</u> : Suppose $\Omega_\pm$ exist and are complete. Also assume that $D(H) = D(K_o) \subseteq D(V)$. Then

$$S = I + \underset{\varepsilon \to +0}{\text{s-lim}}\ \underset{\delta \to +0}{\text{s-lim}} \int_{\mathbb{R}} (R^o_{\lambda - i\varepsilon} - R^o_{\lambda + i\varepsilon})(V - VR_{\lambda + i\delta} V)\, dE^o_\lambda \quad .$$

<u>Remark</u> :

(i) Notice that $V R_{\lambda+i\delta} \in \mathcal{B}(\mathcal{H})$ and $V E^o_\Delta \in \mathcal{B}(\mathcal{H})$ for each compact Δ.

(ii) The order of the limits can in general not be interchanged.

<u>Proof</u> :

(i) Define $\Omega_-^\delta = \delta \int_{-\infty}^0 e^{\delta s} e^{iHs} e^{-iK_o s} ds$, with $\delta > 0$. By Proposition 5.2 :

$$\|\Omega_-^\delta\| \leq \delta \int_{-\infty}^0 ds\, e^{\delta s} \|e^{iHs} e^{-iK_o s}\| = \delta \int_{-\infty}^0 e^{\delta s} ds = 1 \quad . \quad (13.4)$$

Also by Proposition 6.4,

$$\operatorname*{s-lim}_{\delta\to+0} \Omega_-^{\delta} = \Omega_- \equiv \Omega_-^{0} .$$

Let $\varepsilon > 0$ and $I_\varepsilon(\delta) = \int_0^\infty e^{-\varepsilon t}\, e^{iK_o t}\, \Omega_-^{\delta}\, e^{-iK_o t}\, dt$.

The integral exists for each $\delta \geq 0$ by Proposition 5.2(c). One has

$$\| [I_\varepsilon(0) - I_\varepsilon(\delta)] f \| \leq \int_0^\infty dt\, e^{-\varepsilon t}\, \| (\Omega_- - \Omega_-^{\delta})\, e^{-iK_o t} f \| .$$

As $\delta \to 0$, the integrand converges to zero for each t. Also it is majorized uniformly in δ by $2\, e^{-\varepsilon t}\, \|f\| \in L^1(0,\infty)$. Thus by the Lebesgue dominated convergence theorem

$$\operatorname*{s-lim}_{\delta\to+0} I_\varepsilon(\delta) = I_\varepsilon(0) . \tag{13.5}$$

Consequently
$$\operatorname*{s-lim}_{\delta\to+0} \delta\, I_\varepsilon(\delta) = 0 . \tag{13.6}$$

(ii) $$\operatorname*{s-lim}_{\varepsilon\to+0} \operatorname*{s-lim}_{\delta\to+0} (\varepsilon+\delta)\, \delta \int_0^\infty dt\, e^{-\varepsilon t}\, e^{iK_o t} \int_{-\infty}^0 ds\, e^{\delta s}\, e^{iHs}\, e^{-iK_o(s+t)}$$

$$= \operatorname*{s-lim}_{\varepsilon\to+0} \operatorname*{s-lim}_{\delta\to+0} (\varepsilon+\delta)\, I_\varepsilon(\delta) = \operatorname*{s-lim}_{\varepsilon\to+0} \varepsilon\, I_\varepsilon(0)$$

$$= \operatorname*{s-lim}_{\varepsilon\to+0} \varepsilon \int_0^\infty e^{-\varepsilon t}\, e^{iK_o t}\, \Omega_-\, e^{-iK_o t}\, dt$$

$$= \operatorname*{s-lim}_{\varepsilon\to+0} \varepsilon \int_0^\infty e^{-\varepsilon t}\, e^{iK_o t}\, e^{-iHt}\, \Omega_-\, dt = \Omega_+^*\, \Omega_- = S ,$$

where we have used (13.5), (13.6), and Proposition 6.2 and 6.4 and the fact that $F_+ = F_-$ (since $\Omega_\pm$ are complete).

Similarly one obtains

$$\operatorname*{s-lim}_{\varepsilon\to+0} \operatorname*{s-lim}_{\delta\to+0} (\varepsilon-\delta)\, \delta \int_{-\infty}^0 dt\, e^{\varepsilon t}\, e^{iK_o t} \int_{-\infty}^0 ds\, e^{\delta s}\, e^{iHs}\, e^{-iK_o(s+t)}$$

$$= \Omega_-^*\, \Omega_- = I .$$

(iii) As in Proposition 5.6 one obtains

$$\int_0^{\infty} dt\, e^{-\varepsilon t}\, e^{iK_0 t} \int_{-\infty}^{0} ds\, e^{\delta s}\, e^{iHs}\, e^{-iK_0(s+t)} = \int_{\mathbb{R}} R^o_{\lambda-i\varepsilon} R_{\lambda+i\delta}\, dE^o_\lambda$$

$$\int_{-\infty}^{0} dt\, e^{\varepsilon t}\, e^{iK_0 t} \int_{-\infty}^{0} ds\, e^{\delta s}\, e^{iHs}\, e^{-iK_0(s+t)} = - \int_{\mathbb{R}} R^o_{\lambda+i\varepsilon}\, R_{\lambda+i\delta}\, dE^o_\lambda \ .$$

By inserting this in the equations of (ii) :

$$S - I = \underset{\varepsilon\to+0}{\text{s-lim}}\ \underset{\delta\to+0}{\text{s-lim}}\ \delta \int_{\mathbb{R}} \chi_\lambda(\varepsilon,\delta)\, dE^o_\lambda$$

with $\chi_\lambda(\varepsilon,\delta) = (\varepsilon+\delta)\, R^o_{\lambda-i\varepsilon}\, R_{\lambda+i\delta} + (\varepsilon-\delta)\, R^o_{\lambda+i\varepsilon}\, R_{\lambda+i\delta}$. By using the first and the second resolvent equation, this can be transformed into [Problem [13.2]]:

$$\chi_\lambda(\varepsilon,\delta) = i(R^o_{\lambda-i\varepsilon} - R^o_{\lambda+i\varepsilon}) - i(R^o_{\lambda-i\varepsilon} - R^o_{\lambda+i\varepsilon})\, V\, R_{\lambda+i\delta}. \quad (13.7)$$

Now by Proposition 5.6 :

$$i \int R^o_{\lambda\mp i\varepsilon}\, dE^o_\lambda = \int_0^{\pm\infty} dt\, e^{\mp\varepsilon t}\, e^{iK_0 t}\, e^{-iK_0 t} = \pm\varepsilon^{-1} \ .$$

Hence
$$\underset{\delta\to+0}{\text{s-lim}}\ \delta \int (R^o_{\lambda-i\varepsilon} - R^o_{\lambda+i\varepsilon})\, dE^o_\lambda = 0 \quad ,$$

i.e. we need only consider the second term in (13.7). By using Lemma 5.4, we then obtain the desired result :

$$S - I = \underset{\varepsilon\to+0}{\text{s-lim}}\ \underset{\delta\to+0}{\text{s-lim}}\ (-i\delta) \int_{\mathbb{R}} (R^o_{\lambda-i\varepsilon} - R^o_{\lambda+i\varepsilon})\, V\, R_{\lambda+i\delta}\, dE^o_\lambda$$

$$= \underset{\varepsilon\to+0}{\text{s-lim}}\ \underset{\delta\to+0}{\text{s-lim}} \int (R^o_{\lambda-i\varepsilon} - R^o_{\lambda+i\varepsilon})\, V\, R_{\lambda+i\delta}(H-V-\lambda-i\delta)\, dE^o_\lambda$$

$$= \underset{\varepsilon\to+0}{\text{s-lim}}\ \underset{\delta\to+0}{\text{s-lim}} \int (R^o_{\lambda-i\varepsilon} - R^o_{\lambda+i\varepsilon})(V - VR_{\lambda+i\delta}\, V)\, dE^o_\lambda \ .$$

Remark : To apply Lemma 5.4 above, one must show that

$$\| (R^o_{\lambda-i\varepsilon} - R^o_{\lambda+i\varepsilon})\, V\, R_{\lambda+i\delta} \| \le M(\delta,\varepsilon) \quad , \quad \forall\lambda \in \mathbb{R} \ .$$

This is the case (Problem 13.3). ■

We next derive an expression for the so-called *S-matrix* which is closely related to measurable quantities (see later). We have seen that S commutes with $\exp(-iK_0 t)$, hence with every bounded function $\phi(K_0)$ of K_0. This means that S is *decomposable* in the spectral representation of K_0 : let

$$\mathcal{U}_0 f = \{(\mathcal{U}_0 f)_\lambda\} \in L^2([0,\infty), L^2(S^{(n-1)})) .$$

Then $(\mathcal{U}_0 S f)_\lambda = S(\lambda)(\mathcal{U}_0 f)_\lambda$, where for each $\lambda \in (0,\infty)$, $S(\lambda)$ is an operator in $L^2(S^{(n-1)})$, called the *S-matrix at energy* λ.

In other words, in the spectral representation of K_0, S is given by a measurable family $\{S(\lambda)\}$ of operators acting in $L^2(S^{(n-1)})$. [AS]
Since S is unitary, $S(\lambda)$ must be unitary for almost all λ [$S(\lambda)$ is defined for almost all λ], see Problem 13.4.

DEFINITION : $R = S - I$.

$$R(\lambda) = S(\lambda) - I_0 ,$$

where I_0 denotes the identity operator in $L^2(S^{(n-1)})$.
Clearly R is decomposable and $(\mathcal{U}_0 R f) = R(\lambda)(\mathcal{U}_0 f)_\lambda$.

PROPOSITION 13.6 : Let $\mathcal{H} = L^2(\mathbb{R}^n)$, $n \geq 3$, and assume that $V(x) = A(x)\,B(x) = (1 + |x|)^{-\alpha}[B_1(x) + B_2(x)]$ with $\alpha > 1$, $B_1(.) \in L^{p_1}(\mathbb{R}^n) \cap L^{p_2}(\mathbb{R}^n)$, $2 \leq p_1 < n < p_2 \leq \infty$, and $|B_2(x)| \leq c(1 + |x|)^{-\frac{1}{2}-\nu}$, $\nu > 0$. Then

$$R(\lambda) = -2\pi i\, M_A(\lambda)(I + W_{\lambda+io})^{-1} M_B(\lambda)^*$$

for each $\lambda \in (0,\infty)\backslash\Gamma_0$.

Proof : $\Omega_\pm$ exist and are complete by Proposition 8.11. Let Δ be a compact interval in $(0,\infty)\backslash\Gamma_0$ and $f,g \in \mathcal{S}(\mathbb{R}^n)$. Using Proposition 13.5, (7.22), the fact that $R_z^{0*} = R_{\bar{z}}^0$ and Proposition 8.4 (d), we obtain

$$(f, RE^0_\Delta g) = \lim_{\varepsilon\to+0}\lim_{\delta\to+0}\int_\Delta (f, (R^o_{\lambda-i\varepsilon} - R^o_{\lambda+i\varepsilon})\, A(I - BR_{\lambda+i\delta}A)\, B\, dE^0_\lambda g)$$

$$= \lim_{\varepsilon\to+0}\lim_{\delta\to+0}\int_\Delta ((I + W_{\lambda+i\delta})^{-1*} A\, (R^o_{\lambda+i\varepsilon} - R^o_{\lambda-i\varepsilon})\, f, B\, dE^0_\lambda g)$$

$$= \lim_{\varepsilon\to+0}\lim_{\delta\to+0}\int_\Delta (A(R^o_{\lambda+i\varepsilon} - R^o_{\lambda-i\varepsilon})f, (I+W_{\lambda+i\delta})^{-1} M_B(\lambda)^*(\mathcal{U}_0 g)_\lambda)d\lambda.$$

Now by using Lemma 7.5, $\|(I + W_{\lambda+i\delta})^{-1}\| \le M_\Delta < \infty$, $\forall\lambda \in \Delta$, $0 \le \delta \le 1$, and $\|(I + W_{\lambda+i\delta})^{-1} - (I + W_{\lambda+io})^{-1}\| \to 0$ as $\delta \to 0$, $\forall\lambda \in \Delta$. Thus, by the Lebesgue dominated convergence theorem, we may interchange $\lim_{\delta\to+0}$ and $\int_\Delta d\lambda$. By Proposition 8.4(c), the integrand converges as $\varepsilon \to +0$ to

$$- 2\pi i\, (M_A(\lambda)^*\, (\mathcal{U}_0 f)_\lambda, (I+W_{\lambda+io})^{-1}\, M_B(\lambda)^*\, (\mathcal{U}_0 g)_\lambda)\ .$$

Also, by Proposition 5.3 and Lemma 7.7 :

$$\|A\, R^o_{\lambda\pm i\varepsilon}\, f\| = \|A\int_0^{\pm\infty} e^{(i\lambda\mp\varepsilon)t}\, e^{-iK_0 t}\, f\, dt\|$$

$$\le \int_{\mathbb{R}} \|A\, e^{-iK_0 t}\, A\|\, dt\, \|A^{-1}\, f\| < \infty.$$

Using again the Lebesgue dominated convergence theorem, we may interchange $\lim_{\varepsilon\to+0}$ and $\int_\Delta d\lambda$, which leads to

$$(f, RE^0_\Delta g) = - 2\pi i\int_\Delta (M_A(\lambda)^*\, (\mathcal{U}_0 f)_\lambda, (I+W_{\lambda+io})^{-1} M_B(\lambda)^*\, (\mathcal{U}_0 g)_\lambda)\, d\lambda$$

$$= - 2\pi i\int_\Delta ((\mathcal{U}_0 f)_\lambda, M_A(\lambda)(I+W_{\lambda+io})^{-1}\, M_B(\lambda)^*\, (\mathcal{U}_0 g)_\lambda)\, d\lambda.$$

On the other hand

$$(f, RE^0_\Delta g) = \int_\Delta ((\mathcal{U}_0 f)_\lambda, R(\lambda)(\mathcal{U}_0 g)_\lambda)\, d\lambda.$$

Thus there exists a null set $N(f, g) \subset \Delta$ such that

$$((\mathcal{U}_0 f)_\lambda, R(\lambda)(\mathcal{U}_0 g)_\lambda) = -2\pi i\, ((\mathcal{U}_0 f)_\lambda, M_A(\lambda)(I + W_{\lambda+io})^{-1} M_B(\lambda)^* (\mathcal{U}_0 g)_\lambda) \tag{13.8}$$

for all $\lambda \in \Delta\backslash N(f,g)$.

Let $\mathcal{S}_0$ be a countable subset of $\mathcal{S}(\mathbb{R}^n)$ such that for each $\lambda > 0$, the set $\{(\mathcal{U}_0 f)_\lambda \mid f \in \mathcal{S}_0\}$ is dense in $L^2(S^{(n-1)})$. Let $N = \bigcup_{f,g \in \mathcal{S}_0} N(f,g)$. Since $\mathcal{S}_0$ is countable, N is a null set and (13.8) holds for all $f,g \in \mathcal{S}_0$ and $\lambda \in \Delta\backslash N$. Hence

$$R(\lambda) = -2\pi i\, M_A(\lambda)(I + W_{\lambda+io})^{-1} M_B(\lambda)^*, \ \forall \lambda \in \Delta\backslash N. \tag{13.9}$$

Since the right-hand side is defined for all $\lambda \in \Delta$ and the left-hand side almost everywhere, we may assume this identity to hold for all $\lambda \in \Delta$ by possibly changing $R(\lambda)$ on a null set.

$(0,\infty)\backslash\Gamma_0$ is open, hence the union of a countable collection of disjoint open intervals Δ_k. Each $\lambda \in (0,\infty)\backslash\Gamma_0$ belongs to a compact Δ as above, hence $R(\lambda)$ is given by (13.9). ∎

PROPOSITION 13.7 : Assume in Proposition 13.6 that $\alpha > \frac{n}{2}$. Then, for each $\lambda \in (0,\infty)\backslash\Gamma_0$, $R(\lambda)$ is a Hilbert-Schmidt operator, and the function $\lambda \mapsto R(\lambda)$ is continuous in Hilbert-Schmidt norm on $(0,\infty)\backslash\Gamma_0$.

Proof : $M_A(\lambda)$ is a Hilbert-Schmidt operator and $\lambda \mapsto M_A(\lambda)$ is continuous in Hilbert-Schmidt norm, by Proposition 8.1. $M_B(\lambda)^*$ is bounded and $\lambda \mapsto M_B(\lambda)^*$ is continuous in operator norm (Proposition 8.2 and 8.9).
$(I + W_{\lambda+io})^{-1} \in \mathcal{B}(\mathcal{H})$ and $\lambda \mapsto (I + W_{\lambda+io})^{-1}$ is continuous in operator norm. This together with the expression for $R(\lambda)$ given in Proposition 13.6 implies the continuity of $R(\lambda)$ in Hilbert-Schmidt norm. ∎

Since, under the hypotheses of Proposition 13.7, $R(\lambda)$ is a Hilbert-Schmidt operator, it follows from Proposition 2.6 that it is an integral operator in $L^2(S^{(n-1)})$, i.e. of the form

$$[R(\lambda)g]\,(\omega) = \int d\omega'\; r(\lambda\;;\;\omega,\omega')\; g(\omega') \quad , \quad g \in L^2(S^{(n-1)}) \qquad (13.10)$$

with

$$\iint d\omega\; d\omega'\; |r(\lambda\;;\;\omega,\omega')|^2 = \|R(\lambda)\|^2_{HS} < \infty \;. \qquad (13.11)$$

DEFINITION : $f(\lambda\;;\;\omega' \to \omega) \equiv \left(\frac{-2\pi i}{\sqrt{\lambda}}\right)^{\frac{n-1}{2}} \bullet\; r(\lambda\;;\;\omega,\omega')$ is called the *scattering amplitude*.

We shall see later that

$$|f(\lambda\;;\;\omega' \to \omega)|^2$$

is the "probability" for scattering of a particle with incoming velocity $\sqrt{\lambda}\omega'$ into the direction ω.

PROPOSITION 13.8 : Assume that V satisfies (11.10). Then the scattering amplitude is given by

$$f(\lambda\;;\;\omega' \to \omega) = \frac{1}{2}\,(-2\pi i)^{\frac{n+1}{2}}\,\lambda^{\frac{n-3}{4}} \int \overline{\psi^{+}_{\sqrt{\lambda}\omega}(x)}\; V(x)\; \psi^{0}_{\sqrt{\lambda}\omega'}(x)\, dx \;,$$

where the integral exists as a limit in the mean in $L^2(S^{(n-1)})$ for every fixed $\lambda \in (0,\infty)\backslash\Gamma_0$ and $\omega \in S^{(n-1)}$. If in addition $V \in L^1(\mathbb{R}^n)$, the integral exists as a Lebesgue integral.

Proof :

(i) Let $f \in L^2(S^{(n-1)})$. By Proposition 13.6 and 11.1 :

$$R(\lambda)\, f = -2\pi i\; M_A(\lambda)(I + W_{\lambda+io})^{-1}\; M_B(\lambda)^*\, f$$

$$= -2\pi i\; (\mathcal{U}_+\, A\; M_B(\lambda)^*\, f)_\lambda \;.$$

Using now Proposition 11.2(a), we find that

$$[R(\lambda)g](\omega) = -2\pi i\, 2^{-\frac{1}{2}} \lambda^{\frac{(n-2)}{4}} \langle \psi^+_{\sqrt{\lambda}\omega} , A\, M_B(\lambda)^* f \rangle$$

$$= -2\pi i\, 2^{-\frac{1}{2}} \lambda^{\frac{(n-2)}{4}} (A\, \psi^+_{\sqrt{\lambda}\omega} , M_B(\lambda)^* f)$$

$$= -2\pi i\, 2^{-\frac{1}{2}} \lambda^{\frac{(n-2)}{4}} (M_B(\lambda)\, A\, \psi^+_{\sqrt{\lambda}\omega} , f)$$

$$= -2\pi i\, 2^{-\frac{1}{2}} \lambda^{\frac{(n-2)}{4}} \int d\omega'\, \overline{[M_B(\lambda)\, A\, \psi^+_{\sqrt{\lambda}\omega}](\omega')}\, f(\omega') ,$$

since $\phi^+_k = A\, \psi^+_k \in L^2(\mathbb{R}^n)$. Comparison with (13.10) gives

$$r(\lambda\, ;\, \omega,\omega') = -2\pi i\, 2^{-\frac{1}{2}} \lambda^{\frac{(n-2)}{4}} [M_B(\lambda)\, A\, \psi^+_{\sqrt{\lambda}\omega}](\omega') ,$$

i.e.
$$f(\lambda\, ;\, \omega' \to \omega) = (-2\pi i)^{\frac{n+1}{2}}\, 2^{-\frac{1}{2}} \lambda^{-\frac{1}{4}} \overline{[M_B(\lambda)\, A\, \psi^+_{\sqrt{\lambda}\omega}](\omega')} \tag{13.12}$$

(ii) For fixed λ and ω, $f(\lambda\, ;\, .\to \omega) \in L^2(S^{(n-1)})$ and

$$\| f(\lambda\, ;\, . \to \omega)\|_{L^2(S^{(n-1)})} = (2\pi)^{\frac{(n+1)}{2}}\, 2^{-\frac{1}{2}} \lambda^{-\frac{1}{4}} \| M_B(\lambda)\, A\, \psi^+_{\sqrt{\lambda}\omega} \| .$$

Let

$$f_R(\lambda\, ;\, \omega' \to \omega) = (-2\pi i)^{\frac{(n+1)}{2}}\, 2^{-\frac{1}{2}} \lambda^{-\frac{1}{4}} \overline{[M_B(\lambda)\, F_{S_R} A\, \psi^+_{\sqrt{\lambda}\omega}](\omega')} . \tag{13.13}$$

Then

$$\| f(\lambda\, ;\, . \to \omega) - f_R(\lambda\, ;\, . \to \omega)\|_{L^2(S^{(n-1)})}$$
$$\le (2\pi)^{\frac{(n+1)}{2}}\, 2^{-\frac{1}{2}} \lambda^{-\frac{1}{4}} \| M_B(\lambda)\| \, \|(I - F_{S_R})\, A\, \psi^+_{\sqrt{\lambda}\omega}\| ,$$

hence $\underset{R \to \infty}{\text{s-lim}}\, f_R(\lambda\ ;\ . \to \omega) = f(\lambda\ ;\ . \to \omega)$ in $L^2(S^{(n-1)})$. (13.14)

Now from the definition of $M_B(\lambda)$,

$$[M_B(\lambda)\, F_{S_R}\, A\, \psi^+_{\sqrt{\lambda}\omega}](\omega') = 2^{-\frac{1}{2}}\, \lambda^{\frac{(n-2)}{4}} \int_{|x| \le R} dx\, \overline{\psi^0_{\sqrt{\lambda}\omega'}(x)}\, B(x)\, A(x) \psi^+_{\sqrt{\lambda}\omega}(x) \ ,$$

which, together with (13.13) and (13.14), gives the result of the Proposition. Notice that $A(.)\, \psi^+_k(.) \in L^2(\mathbb{R}^n)$ and $B(.)\, \psi^0_{k'}(.) \in L^2_{Loc}(\mathbb{R}^n)$, so that the integrand is an L^1-function on S_R for each $R < \infty$. If $V \in L^1(\mathbb{R}^n)$, then the integrand is in $L^1(\mathbb{R}^n)$ and the limit as $R \to \infty$ exists pointwise, see Proposition 11.8. ∎

COROLLARY 13.9 : If in addition to the hypotheses of Proposition 13.8, $B(.) \in L^2(\mathbb{R}^n)$, then $f(\lambda\ ;\ \omega' \to \omega)$ is continuous in all three variables on $(0,\infty)\backslash\Gamma_0 \times S^{(n-1)} \times S^{(n-1)}$.

Proof : One has

$$f(\lambda\ ;\ \omega' \to \omega) = \frac{1}{2}\, (-2\pi i)^{\frac{(n+1)}{2}}\, \lambda^{\frac{n-3}{4}}\, (A\, \psi^+_{\sqrt{\lambda}\omega},\, B\, \psi^0_{\sqrt{\lambda}\omega'})_{L^2(\mathbb{R}^n)} \ .$$

Now $k \mapsto A\, \psi^+_k$ is continuous in L^2-norm by Proposition 11.3, and so is $k' \mapsto B\, \psi^0_{k'}$ (use the Lebesgue dominated convergence theorem as in the proof of Proposition 11.3). ∎

C. SCATTERING CROSS-SECTION.

A scattering experiment is as follows :
A beam of particles (i.e. a large number of independent particles), all of the same velocity, is scattered by a potential, and one observes the number of particles scattered into different directions. Quantum Mechanics gives a probability distribution for this number, called essentially the scattering cross-section.

The aim of the theory is to calculate the cross-section, assuming that the potential is given.

Precise definition :
Cross-section for a cone(*) C is the number of particles scattered into C per unit time, divided by the number of incoming particles in the beam per unit time and unit surface area (in the hyperplane orthogonal to the velocity of the incoming particles).

Description of the beam :

State of one particle : $g \in L^2(\mathbb{R}^n)$, $\|g\| = 1$.

Hypothesis I - The momentum in the state g is almost sharp : There exists a cone C_0 of small solid angle such that Supp $\tilde{g}(k)$ is contained in a compact subset of C_0 not containing the origin.

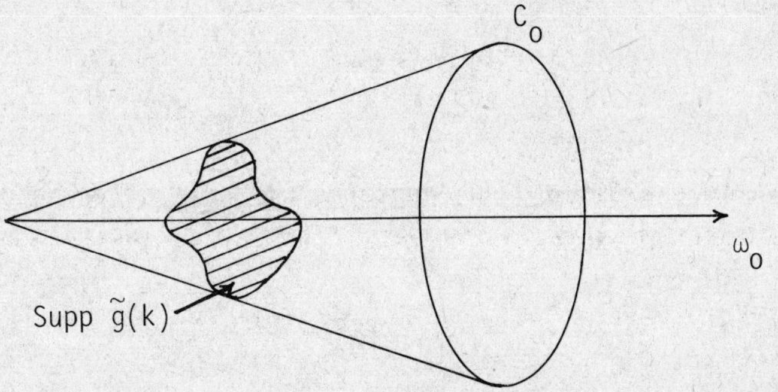

We choose the x_n-axis to lie in the interior of C_0. To describe the other particles in the beam, we translate g by a vector $a \in \mathbb{R}^{n-1}$ in the hyperplane $x_n = 0$. We take a uniform (finally continuous) distribution of the values of a. The translated state is

$$g_a(x) = g(x+a) \quad \text{or} \quad \tilde{g}_a(k) = e^{ik.a}\,\tilde{g}(k) .$$

(*) All cones are assumed to have their apex at the point x = 0.

We denote by $P(g_a, C)$ the probability that the particle in the initial state g_a be scattered into C. We assume that the particles in the beam get scattered independently and there is no multiple scattering. The "number of particles" scattered into C is then simply obtained by adding the individual probabilities, i.e.

$$N(g, C) = \sum_a P(g_a, C) \quad .$$

Let N_0 be the number of points $a \in \mathbb{R}^{n-1}$ lying in the unit square of the hyperplane $x_n = 0$. Then the scattering cross section is

$$\sigma(g, C) = N_0^{-1} N(g, C) = \sum_a \Delta a \; P(g_a, C)$$

where $\Delta a = N_0^{-1}$.

By letting the distribution of a become continuous [i.e. $N_0 \to \infty$], we get

$$\sigma(g, C) = \int_{\mathbb{R}^{n-1}} da \; P(g_a, C) \quad . \tag{13.15}$$

We now have to express $\sigma(g, C)$ in terms of the scattering operator. For this we first obtain an expression for $P(f, C)$ for an arbitrary $f \in L^2(\mathbb{R}^n)$. Define for $t \in \mathbb{R}$, $t \neq 0$:

$$(C_t f)(x) = (2it)^{-\frac{n}{2}} \exp(i \frac{x^2}{4t}) \; \tilde{f}(\frac{x}{2t})$$

$$(Q_t f)(x) = \exp(i \frac{x^2}{4t}) \; f(x) \quad ,$$

where $(2it)^{-\frac{n}{2}} = |2t|^{-\frac{n}{2}} \exp(\mp \frac{in\pi}{4})$ for $t \gtrless 0$ respectively.

Footnote - Fix g. Consider the set of states, i.e. the states $\{g_a\}$ parametrized by $a \in \mathbb{R}^{n-1}$. Each g_a describes a particle of the beam.

LEMMA 13.10 :

(a) C_t and Q_t are unitary operators, and $U_t \equiv \exp(-iK_0 t) = C_t Q_t$.

(b) For each $f \in L^2(\mathbb{R}^n)$:

$$\lim_{t \to \pm\infty} \|U_t f - C_t f\| = 0 \quad .$$

Proof :

(a) Clearly Q_t is unitary. By Lemma 6.10 we have for $f \in L^1(\mathbb{R}^n) \cap L^2(\mathbb{R}^n)$:

$$(U_t f)(x) = (2it)^{-\frac{n}{2}} (2\pi)^{-\frac{n}{2}} \int e^{i\frac{|x-y|^2}{4t}} f(y)\, dy$$

$$= (2it)^{-\frac{n}{2}} e^{i\frac{x^2}{4t}} (2\pi)^{-\frac{n}{2}} \int dy\, e^{-i\frac{x.y}{2t}} (Q_t f)(y)$$

$$= (2it)^{-\frac{n}{2}} e^{i\frac{x^2}{4t}} (\widetilde{Q_t f})(\tfrac{x}{2t}) = (C_t Q_t f)(x) \quad ,$$

i.e. $U_t f = C_t Q_t f$, $\forall f \in L^1(\mathbb{R}^n) \cap L^2(\mathbb{R}^n)$.

Now $f \in L^1 \cap L^2$. This implies that $Q_t^* f \in L^1 \cap L^2$. Hence $U_t Q_t^* f = C_t Q_t Q_t^* f = C_t f$, $\forall f \in L^1 \cap L^2$; thus C_t coincides on $L^1 \cap L^2$ with the unitary operator $U_t Q_t^*$, hence C_t is unitary, and $U_t = C_t Q_t$ by closure.

(b) $\|U_t f - C_t f\|^2 = \|C_t Q_t f - C_t f\|^2 = \|Q_t f - f\|^2$

$$= \int |e^{i\frac{x^2}{4t}} - 1|^2 \, |f(x)|^2 \, dx \to 0 \quad \text{as} \quad t \to \pm\infty,$$

by the Lebesgue dominated convergence theorem. ∎

LEMMA 13.11 : Let C be a cone with apex at the origin, let F_C, G_C be defined as before. Let $\|f\| = 1$ and $P_0(f, C) = \lim_{t \to +\infty} \|F_C\, e^{-iK_0 t} f\|^2$ the probability that the free particle in the state f lies C at $t = +\infty$. Then

$$P_o(f, C) = \|G_C f\|^2 = \int_C dk\, |\tilde{f}(k)|^2 \quad .$$

Remark : This says that $P_o(f, C)$ is just the probability that the momentum of the particle in the state f lies in C, in agreement with intuition.

Proof :

$$| \|F_C U_t f\|^2 - \|F_C C_t f\|^2 |$$

$$= (\|F_C U_t f\| + \|F_C C_t f\|) \, | \|F_C U_t f\| - \|F_C C_t f\| |$$

$$\leq 2\|f\| \, \|F_C U_t f - F_C C_t f\| \to 0 \quad \text{as} \quad t \to \infty \text{ by Lemma 13.10.}$$

Hence $P_o(f, C) = \lim_{t \to \infty} \|F_C U_t f\|^2 = \lim_{t \to \infty} \|F_C C_t f\|^2 = \lim_{t \to \infty} |\frac{1}{2t}|^n \int_C |\tilde{f}(\frac{x}{2t})|^2 \, dx$

$$= \int_C dk\, |\tilde{f}(k)|^2 \quad , \text{ by the change of variable } k = \frac{x}{2t}.$$

(Notice that $x \in C \Leftrightarrow k \in C$ if $t > 0$). ∎

DEFINITION : Let $f \in \mathcal{H}$ and define

$$P(f, C) = \lim_{t \to +\infty} \|F_C\, e^{-iHt}\, \Omega_- f\|^2 \quad .$$

(For $\|f\| = 1$: $P(f, C)$ is the probability that for initial state f the corresponding scattering state $g = \Omega_- f$ lies in C at $t = +\infty$. Here g evolves under the real evolution $\exp(-iHt)$ including the potential).

LEMMA 13.12 :

$$P(f, C) = \|G_C\, S f\|^2 = \int_C dk\, |(\widetilde{Sf})(k)|^2 \; .$$

Remark : This again corresponds to intuition : $P(f, C)$ is the probability that the momentum in the final state Sf lies in C.

Proof :

(i) As above, with the notation $V_t = \exp(-iHt)$:

$$\begin{aligned} \left| \|F_C V_t \Omega_- f\|^2 - \|F_C U_t S f\|^2 \right| & \\ \leq 2\|f\| \, \|F_C V_t \Omega_- f - F_C U_t S f\| & \qquad (13.16) \\ \leq 2\|f\| \, \|\Omega_- f - V_t^* U_t \Omega_+^* \Omega_- f\| . & \end{aligned}$$

The last expression tends to $2\|f\| \, \|\Omega_- f - \Omega_+ \Omega_+^* \Omega_- f\|$ as $t \to \infty$ which is equal to zero since $\Omega_+ \Omega_+^* \Omega_- = E_{ac}(H) \, \Omega_-$ if $\Omega_\pm$ are complete.

(ii) $$P(f, C) = \lim_{t \to \infty} \|F_C V_t \Omega_- f\|^2 = \lim_{t \to \infty} \|F_C U_t S f\|^2$$

$$= P_0(Sf, C) = \|G_C S f\|^2 \text{ by Lemma 13.11.} \quad \blacksquare$$

DEFINITION : $P'(f, C) = \lim_{t \to +\infty} \|F_C(V_t \Omega_- - U_t) f\|^2$.

As above one gets $P'(f, C) = \|G_C(Sf - f)\|^2 = \|G_C R f\|^2$.
$P'(f, C)$ is an auxiliary quantity which will be useful. We have

LEMMA 13.13 :

$$|P'(f, C) - P(f, C)| \leq P_0(f, C) + 2[P_0(f, C) \, P(f, C)]^{\frac{1}{2}} .$$

Proof :

$$|P'(f, C) - P(f, C)| = |\|G_C(Sf - f)\|^2 - \|G_C S f\|^2|$$

$$= |\|G_C S f\|^2 + \|G_C f\|^2 - (G_C S f, G_C f) - (G_C f, G_C S f) - \|G_C S f\|^2|$$

$$\leq \|G_C f\|^2 + 2\|G_C f\| \, \|G_C S f\|$$

$$= P_0(f, C) + 2[P_0(f, C) \, P(f, C)]^{\frac{1}{2}} . \quad \blacksquare$$

We now choose C such that $C \cap C_0 = \{0\}$ (Physically : one observes the scattered particles only outside the direction of the incoming beam).

If g is the initial state with supp $\tilde{g}(k) \in C_0$ as before, then $G_C\, g = G_C G_{C_0}\, g = 0$, i.e.

$$P_0(g, C) = 0 \quad .$$

Similarly Supp $\tilde{g}_a(k) \in C_0$, i.e.

$$P_0(g_a, C) = 0 \quad .$$

It follows from Lemma 13.13 that

$$P'(g_a, C) = P(g_a, C) \quad .$$

Hence by (13.15) and (13.10) :

$$\sigma(g, C) = \int_{\mathbb{R}^{n-1}} da\, P'(g_a, C) = \int da \int_C |(R\tilde{g}_a)(k)|^2\, dk \tag{13.16}$$

$$= \int da \int_{\omega\in C} d\lambda\, d\omega\, \left|\int d\omega'\, r(\lambda\, ;\, \omega, \omega')(\mathcal{U}_0 g)_\lambda\, (\omega')\, e^{i|k|\omega'.a}\right|^2$$

with $|k| = \sqrt{\lambda}$. Set $\phi_\omega(|k|,\omega') \equiv r(\lambda\ ;\ \omega,\omega')(\mathcal{U}_o\, g)_\lambda(\omega')$.

We wish to interchange the order of the first two integrals. For this, we consider

$$I(k) \equiv \int da\ \Big|(2\pi)^{-\frac{(n-1)}{2}} \int d\omega'\ \exp[i \sum_{r=1}^{n-1} k'_r\, a_r]\ \ \phi_\omega(|k|,\omega')\Big|^2$$

where $k' = |k|\ \omega'$.

By the assumption on the support of $\tilde{g}$, we may use instead of $(|k|,\omega')$ the variables $(|k|, k'_1,\dots,k'_{n-1})$. The Jacobian for this change of variables is $|k|^{n-2}\ k'_n$.

Hence, by the Parseval identity in $L^2(\mathbb{R}^{n-1})$:

$$I(k) = \int da\ \Big|(2\pi)^{-\frac{(n-1)}{2}} \int dk'_1 \dots dk'_{n-1}\ \cdot$$

$$\cdot \exp\,[i \sum_{r=1}^{n-1} k'_r\, a_r]\ \phi_\omega(|k|, k'_1,\dots,k'_{n-1})(|k|^{n-2}\ k'_n)^{-1}\Big|^2$$

$$= \int dk'_1 \dots \int dk'_{n-1} |\phi_\omega(|k|, k'_1,\dots,k'_{n-1})|^2\ (|k|^{n-2}\ k'_n)^{-2}$$

$$= \int d\omega' |\phi_\omega(|k|,\omega')|^2\ (|k|^{n-2}\ k'_n)^{-1} \quad .$$

Now $k'_n = |k|\ \cos\theta'$ and $|\cos\theta'|^{-1} \le M' < \infty$ for all $k' \in C_o$. Hence

$$\int d\lambda\ d\omega\ I(\sqrt{\lambda}\omega) \le M' \int d\lambda\ d\omega\ d\omega'\ |k|^{-n+1}\ |r(\lambda\ ;\ \omega,\omega')|^2\ |(\mathcal{U}_o g)_\lambda(\omega')|^2$$

$$\le M^2\ M' \int_\Delta d\lambda\ \lambda^{-\frac{(n-1)}{2}}\ \|R(\lambda)\|^2_{HS} \quad , \qquad (13.17)$$

where we have made the following additional hypothesis :

<u>Hypothesis II</u> -"$|(\mathcal{U}_o g)_\lambda\ (\omega')| \le M\ \ \forall\lambda,\ \omega'$." Δ is such that $(\mathcal{U}_o g)_\lambda\ (\omega) = 0$ whenever $\lambda \notin \Delta$. The last integral is finite provided that $\Delta \subset (0,\infty)\backslash\Gamma_o$,

since $\|R(\lambda)\|_{HS}^2$ is continuous in λ by Proposition 13.7.

Therefore we may apply the Fubini's theorem in (13.16) to interchange the order of the first two integrals, which leads to

$$\sigma(g,C) = (2\pi)^{n-1} \int_C d\lambda d\omega \int d\omega' \, |r(\lambda\ ;\ \omega,\omega')|^2 |(\mathcal{U}_0 g)_\lambda(\omega')|^2 \, \lambda^{-\frac{(n-1)}{2}} \, (\cos\theta')^{-1}$$

$$= \int_C d\lambda d\omega \int d\omega' \, |f(\lambda\ ;\ \omega' \to \omega)|^2 \, |(\mathcal{U}_0 g)_\lambda(\omega')|^2 \, (\cos\theta')^{-1}. \quad (13.18)$$

Since $|(\mathcal{U}_0 g)_\lambda(\omega')|^2$ is the probability that the initial state has energy λ and direction of velocity ω', and since $\cos\theta' \approx 1$ if C_0 is very small, one sees from (13.18) that $|f(\lambda\ ;\ \omega' \to \omega)|^2$ may be interpreted as the probability for scattering at energy λ from the initial direction ω' to the final direction ω. It is called the *differential scattering cross section* at energy λ. The *total scattering cross section* at energy λ is defined to be the integral of $|f(\lambda\ ;\ \omega' \to \omega)|^2$ over all final directions, i.e.

$$\sigma_{tot}(\lambda\ ;\ \omega') = \int_{S^{(n-1)}} d\omega \, |f(\lambda\ ;\ \omega' \to \omega)|^2 . \quad (13.19)$$

If this quantity is averaged over all initial directions ω', one obtains the *averaged total scattering cross section* at energy λ :

$$\sigma_{av}(\lambda) = \frac{1}{a_{n-1}} \int_{S^{(n-1)} \times S^{(n-1)}} d\omega \, d\omega' \, |f(\lambda\ ;\ \omega' \to \omega)|^2 , \quad (13.20)$$

where a_{n-1} denotes the surface area of the sphere S^{n-1}.

Our derivation of Equation (13.18) is valid for example under the hypotheses of Proposition 13.7, which imply that $R(\lambda)$ is Hilbert-Schmidt, in addition to Hypotheses I and II stated above. It is possible to derive (13.18) under considerably weaker assumptions [2], [8]. The remainder of this chapter is devoted to the derivation of a bound on the averaged total scattering cross-section.

We first remark that $\sigma_{av}(\lambda)$ can be simply expressed in terms of the Hilbert-Schmidt norm of $R(\lambda)$:

$$\sigma_{av}(\lambda) = a_{n-1}^{-1} \iint d\omega\, d\omega' \,(2\pi\, \lambda^{-\frac{1}{2}})^{n-1} \,|r(\lambda\, ;\, \omega,\omega')|^2$$

$$= (\frac{2\pi}{\sqrt{\lambda}})^{n-1} \,(a_{n-1})^{-1}\, \|R(\lambda)\|_{HS}^2 \quad . \qquad (13.21)$$

In particular $\sigma_{av}(\lambda)$ is finite if and only if $R(\lambda) \in \mathcal{B}_2$.

Now let $\phi : (0,\infty) \to \mathbb{R}$ be such that $\int_0^\infty \phi(\lambda)^2\, d\lambda = 1$, and define the following operator $P(\phi)$ in $L^2(\mathbb{R}^n)$: [7]

$$(\mathcal{U}_0\, P(\phi)\, f)_\lambda = \phi(\lambda) \int_0^\infty \phi(\mu)(\mathcal{U}_0 f)_\mu\, d\mu, \qquad (13.22)$$

where the integral is a vector-valued integral in $L^2(S^{(n-1)})$. It is easy to check that $P(\phi)$ is an orthogonal projection i.e. $P(\phi)^2 = P(\phi) = P(\phi)^*$. Its range is the subspace $\mathcal{M}(\phi)$ given by

$$\mathcal{M}(\phi) = \{f \in L^2(\mathbb{R}^n) \mid (\mathcal{U}_0 f)_\lambda = \phi(\lambda)\, g\ ,\ g \in L^2(S^{(n-1)})\} \ . \qquad (13.23)$$

<u>LEMMA 13.14</u> : Let ϕ be in $C^1(0,\infty)$, $\phi(\lambda) = 0$ in a neighbourhood of $\lambda = 0$ and $\|\phi\|^2 = \int_0^\infty \phi(\lambda)^2\, d\lambda = 1$. Let W be the multiplication operator in $L^2(\mathbb{R}^n)$ by a function $W(x)$. Then

$$\int_{\mathbb{R}} \|W\, e^{-iK_0 t}\, P(\phi)\|_{HS}^2\ dt = \frac{a_{n-1}}{2}\, (2\pi)^{-n+1}\, \|\lambda^{\frac{(n-2)}{4}}\, \phi\,\|^2\, \|W\|_2^2 \quad (13.24)$$

$$\int_{\mathbb{R}} t^2 \, \| W e^{-iK_0 t} P(\phi)\|^2_{HS} \, dt \tag{13.25}$$

$$= \frac{a_{n-1}}{2} (2\pi)^{-n+1} \left[\| \frac{d}{d\lambda} (\lambda^{\frac{(n-2)}{4}} \phi)\|^2 \, \|W\|^2_2 + \frac{1}{4n} \|\lambda^{\frac{n}{4}-1} \phi\|^2 \, \||x|W\|^2_2 \right. .$$

Proof :

(i) We have $\|W \exp(-iK_0 t) P(\phi)\|_{HS} = \|P(\phi) \exp(iK_0 t) W^*\|_{HS}$. Now $P(\phi) \exp(iK_0 t) W^*$ is an integral operator acting on the Fourier transforms of functions in $L^2(\mathbb{R}^n)$, i.e.

$$[\, F P(\phi) \exp(iK_0 t) W^* f] (k) = \int N(k,k') \tilde{f}(k') \, dk' \quad ,$$

with kernel

$$N(k,k') = (2\pi)^{-\frac{n}{2}} \lambda^{-\frac{(n-2)}{4}} \phi(\lambda) \int_0^\infty \mu^{\frac{(n-2)}{4}} \phi(\mu) \, e^{i\mu t} \, \overline{\tilde{W}(k'-\sqrt{\mu}\omega)} \quad ,(13.26)$$

where $k = \sqrt{\lambda}\omega \in \mathbb{R}^n$. By Proposition 2.6 :

$$\|W e^{-iK_0 t} P(\phi)\|^2_{HS} = \iint |(N(k,k')|^2 \, dk \, dk'$$

$$= \frac{1}{2} \frac{1}{(2\pi)^{n-1}} \int d\lambda \, d\omega \, dk' \, \phi(\lambda)^2 | \frac{1}{\sqrt{2\pi}} \int_0^\infty \mu^{\frac{(n-2)}{4}} \phi(\mu) \, e^{i\mu t} \, \cdot$$

$$\cdot \, \overline{\tilde{W}(k' - \sqrt{\mu}\omega) \, d\mu}|^2 \quad .$$

The integral over $d\lambda$ is 1. We next integrate the above expression w.r.t. dt over the real axis and apply the unitarity of the Fourier transformation to get

$$\int_{\mathbb{R}} \|W e^{-iK_0 t} P(\phi)\|^2 \, dt = \frac{1}{2} \frac{1}{(2\pi)^{n-1}} \int d\omega \, dk' \int_0^\infty d\mu \, |\mu^{\frac{(n-2)}{4}} \phi(\mu) \, \tilde{W}(k'-\sqrt{\mu}\omega)|^2$$

$$= \frac{1}{2} \frac{1}{(2\pi)^{n-1}} \int d\omega \int_0^\infty d\mu \, \mu^{\frac{(n-2)}{2}} \phi(\mu)^2 \, \|\tilde{W}\|_2^2$$

$$= \frac{1}{2} \frac{1}{(2\pi)^{n-1}} a_{n-1} \|\mu^{\frac{(n-2)}{4}} \phi\|^2 \, \|W\|_2^2 \quad ,$$

which proves (13.24).

(ii) The operator $t\,P(\phi)\exp(iK_0 t)\,W^*$ has kernel $t\,N(k,k')$. By using the identity $t\exp(i\mu t) = -i\frac{d}{d\mu}\exp(i\mu t)$ and integrating by parts in the integral in (13.26), we get

$$t\,N(k,k') = (2\pi)^{-\frac{n}{2}} \lambda^{-\frac{(n-2)}{4}} \phi(\lambda)\, i \,\cdot$$

$$\cdot \int_0^\infty d\mu \, e^{i\mu t} \, \overline{\{\tilde{W}(k' - \sqrt{\mu}\omega)} \frac{d}{d\mu}(\mu^{\frac{(n-2)}{4}} \phi(\mu)) +$$

$$+ \mu^{\frac{(n-2)}{4}} \phi(\mu) \frac{d}{d\mu} \overline{\tilde{W}(k' - \sqrt{\mu}\omega)}\} \; .$$

We set $Z_j(x) = x_j\,W(x)$, $j = 1,\dots,n$ and $\xi = \sqrt{\mu}\omega$. Then

$$\frac{d\tilde{W}(k'-\sqrt{\mu}\omega)}{d\mu} = \frac{1}{2}\mu^{-\frac{1}{2}} \sum_{j=1}^{n} \frac{\xi_j}{|\xi|} \frac{\partial \tilde{W}(k'-\xi)}{\partial \xi_j}$$

$$= \frac{i}{2}\mu^{-\frac{1}{2}} \omega.\tilde{Z}(k'-\sqrt{\mu}\omega).$$

Hence, as in (i)

$$\int_{\mathbb{R}} t^2 \| W e^{-iK_0 t} P(\phi)\|^2_{HS} = \frac{1}{2} \frac{1}{(2\pi)^{n-1}} \int d\omega\, dk' \int_0^\infty d\mu\, |\widetilde{W}(k'-\sqrt{\mu}\omega) \frac{d}{d\mu}(\mu^{\frac{(n-2)}{4}} \phi(\mu))$$

$$- \frac{i}{2} \mu^{\frac{n}{4}-1} \phi(\mu)\omega \,.\, \widetilde{Z}(k' - \sqrt{\mu}\omega)|^2$$

$$= \frac{1}{2} \frac{1}{(2\pi)^{n-1}} \int_0^\infty d\mu \int d\omega\, \|\alpha\widetilde{W}(.) - i\beta\omega.\widetilde{Z}(.)\|^2_{L^2(\mathbb{R}^n)} \tag{13.27}$$

where $\alpha = \frac{d}{d\mu} [\mu^{\frac{(n-2)}{4}} \phi(\mu)]$ and $\beta = \frac{1}{2} \mu^{\frac{n}{4}-1} \phi(\mu)$.

Now for $\alpha,\beta \in \mathbb{R}$:

$$\int_{S^{(n-1)}} d\omega\, \|\alpha\widetilde{W}(.) - i\beta\omega.\widetilde{Z}(.)\|^2_{L^2(\mathbb{R}^n)} = a_{n-1}(\alpha^2 \|W\|_2^2 + \frac{\beta^2}{n} \sum_{j=1}^{n} \|Z_j\|^2) .$$

Upon inserting this into (13.27), one obtains (13.25). ■

<u>PROPOSITION 13.15</u> : Let ϕ be as in Lemma 13.14. Then for each $\varepsilon \in (0,\frac{1}{2}]$:

$$\int_0^\infty \lambda^{\frac{(n-1)}{2}} \sigma_{av}(\lambda)\, \phi(\lambda)^2\, d\lambda \le C(\phi,\varepsilon)\, \|(1 + |x|)^{\frac{1}{2}+\varepsilon} V\|_2^2 \tag{13.28}$$

with

$$C(\phi,\varepsilon) = 4^{\varepsilon-\frac{1}{2}} \frac{\Gamma(\varepsilon)\Gamma(1+\varepsilon)}{\Gamma(1+2\varepsilon)} \|\lambda^{\frac{(n-2)}{4}} \phi\|^{1-2\varepsilon} \cdot$$

$$\cdot \left[\|\lambda^{\frac{(n-2)}{4}} \phi\|^2 + \|\frac{d}{d\lambda} (\lambda^{\frac{(n-2)}{4}} \phi)\|^2 + \frac{1}{4n} \|\lambda^{\frac{n}{4}-1} \phi\|^2 \right]^{\frac{1}{2}+\varepsilon}$$

Remark : (13.28) implies in particular that $\sigma_{av}(\lambda) < \infty$ a.e. if $(1 + |x|)^{\frac{1}{2}+\varepsilon} V(x) \in L^2(\mathbb{R}^n)$ for some $\varepsilon > 0$.

Proof :

(i) Let $\{e_k\}_{k=1}^{\infty}$ be an orthonormal basis of $L^2(S^{(n-1)})$. Define g_k by $(\mathcal{U}_0 \, g_k)_\lambda = \phi(\lambda) \, e_k$. Then $\{g_k\}_{k=1}^{\infty}$ is an orthonormal basis of the subspace $\mathcal{M}(\phi)$. Hence

$$\|R\, P(\phi)\|_{HS}^2 = \sum_k \|R\, g_k\|^2 = \sum_k \int_0^\infty d\lambda \, \|R(\lambda)\, \phi(\lambda)\, e_k\|^2$$

$$= \int_0^\infty d\lambda \, \phi(\lambda)^2 \, \|R(\lambda)\|_{HS}^2 = \frac{a_{n-1}}{(2\pi)^{n-1}} \int_0^\infty \lambda^{\frac{(n-1)}{2}} \phi(\lambda)^2 \, \sigma_{av}(\lambda) \, d\lambda . \tag{13.29}$$

(ii) Using the definition and properties of the wave operators, one finds that

$$\|R\, P(\phi)\|_{HS} = \|(S - I)\, P(\phi)\|_{HS} = \|\Omega_+^*(\Omega_- - \Omega_+)\, P(\phi)\|_{HS}$$

$$\leq \|(\Omega_+ - \Omega_-)\, P(\phi)\|_{HS} = \left\| \int_{-\infty}^{\infty} \frac{d}{dt} e^{iHt} e^{-iK_0 t} P(\phi)\, dt \right\|_{HS}$$

$$= \left\| \int_{-\infty}^{\infty} e^{iHt} \, V \, e^{-iK_0 t} P(\phi)\, dt \right\|_{HS}$$

$$\leq \int_{-\infty}^{+\infty} \|V\, e^{-iK_0 t} P(\phi)\|_{HS} \, dt$$

$$\leq \left[\int_{-\infty}^{+\infty} (1+t^2)^{-(\frac{1}{2}+\varepsilon)} dt \int_{-\infty}^{+\infty} (1+t^2)^{\frac{1}{2}+\varepsilon} \|V\, e^{-iK_0 t} P(\phi)\|_{HS}^2 \, dt \right]^{\frac{1}{2}}$$

$$= \left[4^{\varepsilon} \frac{\Gamma(\varepsilon)\Gamma(1+\varepsilon)}{\Gamma(1+2\varepsilon)} \int_{-\infty}^{+\infty} (1+t^2)^{\frac{1}{2}+\varepsilon} \|V\, e^{-iK_0 t} P(\phi)\|_{HS}^2 \, dt \right]^{\frac{1}{2}} .$$

Hence, by (13.29) ,

$$\int_0^{\infty} \lambda^{\frac{(n-1)}{2}} \phi(\lambda)^2 \sigma_{av}(\lambda)\, d\lambda \le \frac{(2\pi)^{n-1}}{a_{n-1}} 4^{\varepsilon} \frac{\Gamma(\varepsilon)\Gamma(1+\varepsilon)}{\Gamma(1+2\varepsilon)} \cdot$$

$$\cdot \int_{-\infty}^{+\infty} (1+t^2)^{\frac{1}{2}+\varepsilon} \| V e^{-iK_0 t} P(\phi) \|_{HS}^2 \, dt. \qquad (13.30)$$

(iii) To estimate the integral over dt in (13.30), we set $W(x) = (1+|x|)^{\frac{1}{2}+\varepsilon} V(x)$ and denote $M_t(x,y)$ the kernel of the Hilbert-Schmidt operator $W \exp(-iK_0 t)\, P(\phi)$ (see Proposition 2.6). Then the integral over dt in (13.30) is nothing but the norm in $L^1(\mathbb{R} \times \mathbb{R}^n \times \mathbb{R}^n ;\, dt\, dx\, dy)$ of the function

$$(1+t^2)^{\frac{1}{2}+\varepsilon} (1+|x|)^{-2(\frac{1}{2}+\varepsilon)} |M_t(x,y)|^2$$

$$= \left[(1+t^2)^{\frac{1}{2}+\varepsilon} (1+|x|)^{-2(\frac{1}{2}+\varepsilon)} |M_t(x,y)|^{2(\frac{1}{2}+\varepsilon)}\right] [|M_t(x,y)|^{2(\frac{1}{2}-\varepsilon)}].$$

We set $p = (\frac{1}{2}+\varepsilon)^{-1}$, $q = (\frac{1}{2}-\varepsilon)^{-1}$ and notice that $p^{-1} + q^{-1} = 1$. Hence by the Hölder inequality :

$$\int_{-\infty}^{+\infty} (1+t^2)^{\frac{1}{2}+\varepsilon} \| V e^{-iK_0 t} P(\phi) \|_{HS}^2 \, dt$$

$$\le \left(\int dt\, dx\, dy\, (1+t^2)(1+|x|)^{-2} |M_t(x,y)|^2\right)^{\frac{1}{p}} \cdot$$

$$\cdot \left(\int dt\, dx\, dy\, |M_t(x,y)|^2\right)^{\frac{1}{q}}$$

$$= \left[\int dt\, (1+t^2)\, \| (1+|Q|)^{-1} W e^{-iK_0 t} P(\phi) \|_{HS}^2\right]^{\frac{1}{2}+\varepsilon} \cdot$$

$$\cdot \left(\int dt\, \| W e^{-iK_0 t} P(\phi) \|_{HS}^2 \right)^{\frac{1}{2}-\varepsilon} .$$

The last two integrals can easily be estimated by Lemma 13.14, and upon insertion of these estimates into (13.30) one obtains (13.28). ■

D - THE BORN APPROXIMATION AND AN INVERSE PROBLEM.

In this section, we consider a first order approximation to the eigenfunctions $\psi^{\pm}_{\sqrt{\lambda}\omega}(x)$ and the scattering amplitude (see Proposition 13.8) :

$$f(\lambda\ ;\ \omega' \to \omega) = \frac{1}{2}(-2\pi i)^{\frac{n+1}{2}} \lambda^{\frac{n-3}{4}} \int \overline{\psi^{+}_{\sqrt{\lambda}\omega}(x)}\, V(x)\, \psi^{0}_{\sqrt{\lambda}\omega'}(x)\, dx \qquad (13.31)$$

called the Born approximation, showing that, at least for $V \in L^1(\mathbb{R}^n)$ which satisfies a further technical condition we give later, one has

$$f(\lambda\ ;\ \omega' \to \omega) = \frac{1}{2}(-i)^{\frac{n+1}{2}} \sqrt{2\pi}\, \lambda^{\frac{n-3}{4}} [\tilde{V}(\sqrt{\lambda}(\omega-\omega')) + R(\lambda,\omega,\omega')], \qquad (13.32)$$

where

$$\sup_{\omega,\omega'} |R(\lambda,\omega,\omega')| \to 0 \quad \text{as } \lambda \to \infty. \qquad (13.33)$$

To begin, we have the Lippmann-Schwinger equation derived in (11.8) :

$$\psi^{\pm}_{k} = \psi^{0}_{k} - (K_0 - \lambda \pm io)^{-1} V \psi^{\pm}_{k} \quad \text{with } \lambda = |k|^2\ ;\ \psi^{0}_{k} = (2\pi)^{-\frac{n}{2}} e^{ik.x}. \qquad (13.34)$$

Equivalently, with $\phi^{\pm}_{k} = A\,\psi^{\pm}_{k}$, $\phi^{0}_{k} = (2\pi)^{-\frac{n}{2}} A\, e^{ik.x}$, one has (see (11.11)) :

$$(I + W^{*}_{\lambda\pm io})\, \phi^{\pm}_{k} = \phi^{0}_{k}\ , \qquad (13.35)$$

where $W_{\lambda\pm io} = B(K_0 - \lambda \mp io)A$ and $V = AB$. We will suppose that the potential V is such that

$$\|W_{\lambda\pm io}\| \to 0 \text{ as } |\lambda| \to \infty. \tag{13.36}$$

Later we will give a more explicit condition on V which guarantees that (13.36) is satisfied. Wanted this condition, we see that, if

$$A \in L^2(\mathbb{R}^n)$$

then, for $|k|$ large, there is a unique solution $\phi_k^{\pm}$ to (13.35) :

$$\phi_k^{\pm} = (2\pi)^{-\frac{n}{2}} (I + W^*_{\lambda\pm io})^{-1}(A\, e^{ik.x}). \tag{13.37}$$

In particular

$$\phi_k^{\pm} = \phi_k^o + r_k^{\pm} \tag{13.38}$$

with

$$r_k^{\pm} = - W^*_{\lambda\pm io}(I + W^*_{\lambda\pm io})^{-1}\ \phi_k^o ,$$

so that by (13.36),

$$\sup_{\omega} \|r_k^{\pm}\|_2 \to 0 \quad \text{as} \quad |k| \to \infty. \tag{13.39}$$

We now show that this leads to the following result, the Born approximation for the scattering amplitude.

<u>THEOREM 13.16</u> : Suppose $V = AB$, $A,B \in L^2(\mathbb{R}^n)$, and assume (13.36) holds. Then the Born approximation (13.32) holds, with the estimate (13.33).

<u>Proof</u> : By the integral formula for $f(\lambda\,;\omega \to \omega')$, we have (13.32) with

$$R(\lambda,\omega,\omega') = \int e^{ik'.x}\ \overline{r_k^+(x)}\ B(x)\ dx .$$

Thus

$$\sup_{\omega,\omega'} |R(\lambda,\omega,\omega')| \leq \sup_{\omega} \|B\, r_k^+\|_1$$

$$\leq \sup_{\omega} \|B\|_2\, \|r_k^+\|_2 \quad ,$$

so the estimate (13.39) completes the proof. ■

A simple corollary is the following inverse result

COROLLARY 13.17 : Assume $V \in L^1(\mathbb{R}^n)$ and that (13.36) holds. Then the behaviour of $f(\lambda\,;\,\omega' \to \omega)$ for large λ uniquely determines V.

Proof : Indeed

$$2\,\lambda^{-\frac{n-3}{4}}\, f(\lambda\,;\,\omega' \to \omega) = (-i)^{\frac{n+1}{2}}\, \sqrt{2\pi}\, [\tilde{V}(\sqrt{\lambda}(\omega-\omega')) + R(\lambda,\omega,\omega')]$$

with R satisfying the estimate (13.33). Hence, picking $\omega_\lambda, \omega'_\lambda \in S^{n-1}$ such that $\sqrt{\lambda}(\omega_\lambda - \omega'_\lambda) = \xi$, for some fixed $\xi \in \mathbf{R}^n$, we see that

$$\lim_{\lambda\to\infty} \lambda^{-\frac{n-3}{4}}\, f(\lambda\,;\,\omega'_\lambda \to \omega_\lambda) = \frac{1}{2}\sqrt{2\pi}\,(-i)^{\frac{n+1}{2}}\, \tilde{V}(\xi)\ .$$

Consequently, the behaviour of $f(\lambda\,;\,\omega'_\lambda \to \omega_\lambda)$ for *large* λ, *near* the forward direction $\omega' = \omega$ but not quite *on* the forward direction determines $\tilde{V}(\xi)$ for each $\xi \in \mathbb{R}^n$, and hence V(x) is uniquely determined. ■

This result was remarked by (R.G. Newton, [30]), who also emphasized that it is not of great practical value, since experimental determination of the amplitude so near the forward direction would require too much precision of measurement. See (R.G. Newton, [30]) for other approaches to the inverse scattering problem in $\mathbb{R}^3$. Also (Deift and Trubowitz, [17]) have shown that *backscattering* alone uniquely determines a class of potentials in $\mathbb{R}^3$, at least locally.

We now give a condition on the potential V which guarantees (13.36).

LEMMA 13.18 : Let $n \geq 3$, and let $\phi,\psi : \mathbb{R}^n \to \mathbb{C}$ belong to $L^p(\mathbb{R}^n) \cap L^q(\mathbb{R}^n)$ for some $2 \leq p < n < q < \infty$. Let S,T be the multiplication operators in $L^2(\mathbb{R}^n)$ by ϕ,ψ respectively, and set

$$Z_{\lambda+i\varepsilon} = i \int_0^\infty dt\, e^{i\lambda t-\varepsilon t}\, S\, e^{-iK_0 t}\, T\,.$$

Then, for each $\varepsilon \geq 0$, one has

$$\lim_{\lambda\to\infty} \|Z_{\lambda+i\varepsilon}\| = 0\,.$$

Proof : By Lemma 7.7, we have $\|S\, e^{-iK_0 t}\, T\| \in L^1(\mathbb{R})$ as a function of t, and $Z_{\lambda+i\varepsilon}$ is norm continuous on [0,1] as a function of ε. Thus $Z_{\lambda+i\varepsilon}$ is the inverse Fourier transform of the function $t \mapsto e^{-\varepsilon t}\, \chi_{[0,\infty)}(t)\, S\, e^{-iK_0 t}\, T$ which belongs to $L^1(\mathbb{R}, \mathcal{B}(\mathcal{H})\, ; dt)$. It then follows from the Riemann-Lebesgue lemma applied to Banach space-valued functions (See L. Schwartz, [S] vol. II, Complements Theorem 1) that $\|Z_{\lambda+i\varepsilon}\| \to 0$ as $\lambda \to \infty$. ∎

PROPOSITION 13.19 : Let $\mathcal{H} = L^2(\mathbb{R}^n)$, $n \geq 3$, and assume that $V(x) = A(x)\, B(x) \equiv (1 + |x|)^{-\alpha}\, B(x)$ with $\alpha > \frac{n}{2}$ and $B(.) \in L^2(\mathbb{R}^n) \cap L^q(\mathbb{R}^n)$ for some $q \in (n,\infty)$. Then the conclusions of Proposition 13.16 and Corollary 13.17 hold.

Proof : It suffices to remark that $W_{\lambda\pm io} = B(K_0 - \lambda\mp io)^{-1} A$ satisfies (13.36), by Lemma 13.18. ∎

Problems

1.1. : In the example of Chapter 1, verify that

$$\Omega_{\pm} = \lim_{t \to \pm\infty} V_{-t} U_t ,$$

exist as strong limits.

1.2. : Verify that $F\Omega_{\pm}^{-1}$ diagonalize the operator H.

1.3. : Define for $t \in \mathbb{R}$:

$$(U_t f)(x) = f(x - t) .$$

Verify that (i) U_t is unitary.

(ii) $\{U_t\}$ forms a group in t.

(iii) $\{U_t\}$ is strongly continuous.

2.1. : Let T be a finite rank operator. Prove the following statements :

(a) T^* is in $\mathcal{B}_F$.

(b) If $B \in \mathcal{B}(\mathcal{H})$, then TB and BT belong to $\mathcal{B}_F$.

(c) If T_1 is of finite rank, then so is $T + \alpha T_1$, $(\alpha \in \mathbb{C})$.

2.2. : Prove Lemma 2.4.

2.3. : Show the null space of any closed densely defined operator is a subspace of $\mathcal{H}$ and is the orthogonal complement of the range of its adjoint.

2.4. : Prove Lemma 2.5 (a) and (c).

2.5. : Let A be closable, $\bar{A}$ its closure and A' a closed extension of A. Prove that $\bar{A} \subseteq A'$.

2.6. : Prove the *Polarization identity*

$$4(f,g) = \|f + g\|^2 - \|f - g\|^2 - i\|f + ig\|^2 + i\,\|f - ig\|^2 .$$

2.7. : Give an example of a compact operator which is not in the Hilbert-Schmidt class.

2.8. : Show that $A^*A \in \mathcal{B}_\infty$ if and only if $A \in \mathcal{B}_\infty$.

2.9. : Let $H = L^2(\mathbb{R})$ and let Q be the maximal multiplication operator defined by

$$(Qf)(x) = x\, f(x) .$$

Show that $(I - Q)^{-1}$ exists but is unbounded.

3.1. : Prove that a symmetric operator A in $\mathcal{H}$ is essentially self-adjoint if and only if the range of both of the operators $A \pm i$ is dense in $\mathcal{H}$.

3.2. : Prove the uniform continuity of Lemma 3.5.

3.3. : Give an example where $A = A^*$, $B = B^*$, B is A-bounded with A-bound $\beta_0 = 1$ and A+B is not self-adjoint.

3.4. : Let $A = A^*$, B symmetric and

$$\|Bf\| \leq \alpha'^2\, \|f\|^2 + \|Af\|^2 \quad \text{for all} \quad f \in D(A) ,$$

[i.e. (3.7) holds with $\beta' = 1$] . Show that A+B is essentially self-adjoint on D(A).

3.5. : Let $A \in \mathcal{B}(\mathcal{H})$ and (f,Af) real for all $f \in \mathcal{H}$. Then A is self-adjoint.

3.6. : Classify the self-adjoint extensions of $-\frac{d^2}{dx^2}$ on $C_0^\infty(0,2\pi)$ and interpret in terms of scattering on a circle with a distinguished point.

3.7. : Show that, if A self-adjoint, then $(A+i)(A+in)^{-1}$ belongs to $\mathcal{B}(\mathcal{H})$ for every integer $n \geq 1$ and $\{(A+i)(A+in)^{-1}\}$ converges strongly to zero as $n \to \infty$.

4.1. : Prove that for each $\lambda \in \mathbb{R}$, $\underset{\varepsilon \to +0}{\text{s-lim}}\, E_{\lambda-\varepsilon} \equiv E_{\lambda-0}$ exists and is a projection.

4.2. : Prove Lemma 4.3.

4.3. : Using the Spectral Theorem, show that there is a dense set $\mathcal{D}$ such that $\sum_{n=0}^{\infty} \frac{A^n f}{n!}$ converge strongly for each $f \in \mathcal{D}$ to $\exp(A)\, f$, $\exp(A)$ being defined by the functional calculus.

4.4. : Prove the following statements :

(i) $\lambda \in \sigma_d(A) \Leftrightarrow \lambda$ is isolated in $\sigma(A)$ and $0 < \dim E_{\{\lambda\}}\mathcal{H} < \infty$,

(ii) $\lambda \in \sigma_e(A) \Leftrightarrow \dim E_{(\lambda-\varepsilon,\lambda+\varepsilon]}\mathcal{H} = \infty$, for each $\varepsilon > 0$.

4.5. : If $\lambda \in \sigma(A)$ but $\lambda \notin \sigma_p(A)$, then $A - \lambda$ is invertible and $(A-\lambda)^{-1}$ is densely defined but unbounded.

4.6. : Prove Proposition 4.5.

Let $A = A^*$. Show :

(a) $AR_z \in \mathcal{B}(\mathcal{H})$ for all $z \in \rho(A)$.

(b) The resolvent R_z satisfies the first resolvent equation :

$$R_z - R_{z'} = (z-z')\, R_z R_{z'}, \text{ for } z,z' \in \rho(A).$$

(c) $\rho(A)$ is open in the complex plane and R_z is uniformly holomorphic in each connected component of $\rho(A)$.

4.7. : Let $A = A^*$, $\{E_\lambda\}$ its spectral family. Show that

$$\|A\| = \max\{|m|, |M|\},$$

where m and M are lower and upper bound of $\{E_\lambda\}$.

5.1. : Prove Proposition 5.1.

5.2. : Prove Proposition 5.2.

5.3. : Let $A = A^*$ and $R_z = (A-z)^{-1}$. Show that, if $a, b \notin \sigma_p(A)$, one has

$$E_{(a,b)} = E_{[a,b]} = E_{(a,b]} = \underset{\varepsilon \to +0}{\text{s-lim}}\, (2\pi i)^{-1} \int_a^b d\lambda\, [R_{\lambda+i\varepsilon} - R_{\lambda-i\varepsilon}].$$

5.4. : Prove the Equation (5.24).

5.5. : Suppose that $\int_{\mathbb{R}} A_\lambda \, dE_\lambda$ and $\int_{\mathbb{R}} dE_\lambda \, B_\lambda$ exist. Prove the following statements.

(a) $\int_{\mathbb{R}} A_\lambda \, dE_\lambda \, B_\lambda$ exists.

(b) $\int_{\mathbb{R}} A_\lambda \, dE_\lambda \int_{\mathbb{R}} dE_\mu \, B_\mu = \int_{\mathbb{R}} A_\lambda \, dE_\lambda \, B_\lambda$.

5.6. : Let $\{U_t\}$, $-\infty < t < +\infty$, be a strongly continuous one-parameter unitary group in $\mathcal{H}$. Define a linear operator A (called the *infinitesimal generator* of $\{U_t\}$) as follows :

$$D(A) = \{f \mid \text{s-lim}\, i\tau^{-1} (U_\tau - 1)f \ \text{ exists as } \tau \to 0\}$$

and

$$A(f) = \text{s-lim}\, i\tau^{-1} (U_\tau - I)f \quad \text{for} \quad f \in D(A).$$

Prove

(a) $D(A)$ is dense.

(b) A is self-adjoint.

(c) For any $z \in \mathbb{C}$ ($\text{Im } z \neq 0$), the operator $(A-z)^{-1}$ is bounded and defined everywhere and given by

$$(A-z)^{-1} = \begin{cases} i \int_0^\infty e^{izs} e^{-iAs} \, ds & \text{if} \quad \text{Im } z > 0 \\ -i \int_{-\infty}^0 e^{izs} e^{-iAs} \, ds & \text{if} \quad \text{Im } z < 0 . \end{cases}$$

6.1. : Prove Proposition 6.4.

6.2. : Let $A = A^*$, P be a projection such that $PA \subseteq AP$. Show $PD(A) = D(AP)$.

6.3. : Prove the existence of Ω_- as in Proposition 6.13.

6.4. : Prove Proposition 6.14.

6.5. : Prove the following statements

(a) Let F_r be the projection

$$(F_r f)(x) = \begin{cases} f(x) & |x| \leq r \\ 0 & |x| > r \end{cases}$$

in $L^2(\mathbb{R}^n)$. Show that $\underset{t \to \pm\infty}{\text{s-lim}}\, F_r e^{-iK_0 t} f = 0$ for each $f \in L^2(\mathbb{R}^n)$ and each $r < \infty$.

(b) Let ψ be the multiplication operator in $L^2(\mathbb{R}^n)$ by $\psi(x)$, where $\psi \in L^\infty(\mathbb{R}^n)$ and $\lim_{|x| \to \infty} \psi(x) = 0$. Show that $\underset{t \to \pm\infty}{\text{s-lim}}\, \psi e^{-iK_0 t} f = 0$ for each $f \in L^2(\mathbb{R}^n)$.

6.6. : Suppose $V = V_1 + V_2$ with $V_1 \in L^2(\mathbb{R}^n)$, $V_2(x) = 0$ for $|x| < R$ and $|V_2(x)| \le c|x|^{-\beta}$ for $|x| > R$ and some $0 < \beta \le \frac{n}{2}$. Show that for $f \in \mathcal{S}(\mathbb{R}^n)$ and $|t| > 2$:

$$\|V e^{-iK_0t} f\| \le c_1|t|^{-\beta} \quad \text{if} \quad 0 < \beta < \frac{n}{2}$$

$$\|V e^{-iK_0t} f\| \le c_2|t|^{-\frac{n}{2}} \log |t| \quad \text{if } \beta = \frac{n}{2} .$$

What can you conclude about the existence of the wave operators ?

6.7. : Let $H = H_0 + V$ and V H_0-bounded with H_0-bound $\beta_0 < 1$. Show that for $f \in D(H_0) \cap \mathcal{H}_{ac}(H_0)$:

$$V\Omega_- f = \underset{\varepsilon \to +0}{\text{s-lim}} \int_{\mathbb{R}} [V - V(H-\lambda-i\varepsilon)^{-1}V] \; dE_\lambda^0 .$$

7.1. : Let $n \ge 3$, V a self-adjoint operator of rank one, i.e. $Vf = \gamma(h,f)h$ with $\gamma \in \mathbb{R}$. Assume that $h \in L^1(\mathbb{R}^n) \cap L^2(\mathbb{R}^n)$. Let $H = K_0 + V$. Show that $\Omega_\pm$ exist and are asymptotically complete.

7.2. : Let $H = K_0 + V$ with $V : \mathbb{R}^3 \to \mathbb{R}$, $V \in L^1(\mathbb{R}^3) \cap L^\infty(\mathbb{R}^3)$. Show that the wave operators $\Omega_\pm$ are complete. [Let A be the multiplication by $|V(x)|^{1/2}$, B be the multiplication by $|V(x)|^{1/2}$ sign $V(x)$ and C be the multiplication by $(1 + |x|^2)^{-1}$ as in Proposition 7.6.

7.3. : Let $A = A^*$ and $\Delta \subseteq \mathbb{R}$ an interval. Assume that there exists a dense set $\mathcal{D}$ in $\mathcal{H}$ such that for each $f \in \mathcal{D}$,

$$|f,(A-\lambda-i\varepsilon)^{-1}f)| \le M = M(f) < \infty \quad \text{for all } \lambda \in \Delta,$$

$0 < \varepsilon \le 1$. Show that the part of H in $E_\Delta\mathcal{H}$ is absolutely continuous.

7.4. : (a) H, H_o, A, B be self-adjoint, $A \in \mathcal{B}(\mathcal{H})$, B H_o-bounded and H-bounded and $H = H_o + \gamma AB$ with $D(H) = D(H_o)$ and $\gamma \in \mathbb{R}$. Assume $\|A e^{-iH_o t} A\| \in L^1(\mathbb{R})$, $\|B(H_o - \lambda \mp i\varepsilon)^{-1} A\| \le M_1 < \infty \quad \forall \lambda \in \mathbb{R}$ and $0 < \varepsilon \le 1$, that there exists C with dense range such that $\|B(H_o - \lambda \mp i\varepsilon)^{-1} C\| \le M_2 < \infty$ for all $\lambda \in \mathbb{R}$ and $0 < \varepsilon \le 1$. Assume that H_o is spectrally absolutely continuous and that the wave operators exist. Show that, if $|\lambda| < M_1^{-1}$, the wave operators are unitary, hence they are complete and H is spectrally absolutely continuous.

(b) Determine M_1, for the class of potentials given in Proposition 7.8, with $H_o = K_o$.

8.1. Prove Proposition 8.11.

9.1. : Consider the ordinary differential operator H defined by

$$H = -\frac{d^2}{dr^2} + \gamma \chi_{[0,1]}(r) + \ell(\ell+1)\, r^{-2} \quad \text{in} \quad L^2(0,\infty) ,$$

where $\chi_{[0,1]}$ is the characteristic function of the interval [0,1]. Show that, for certain values of γ, $\lambda = 0$ belongs to Γ_o. Show that, if $\ell = 1,2,\ldots$, then $\lambda = 0$ is an eigenvalue of H_1 whereas for $\ell = 0$ this is not so.

10.1. : Prove that the operator B defined by (10.57) and (10.58) is self-adjoint.

10.2. : Prove (10.59).

10.3. : Prove the estimate (10.25). Evaluate the integral on the r.h.s. of (10.25).

11.1. : Derive the expression for the Green's function $G_z(u)$ given in the equation preceding (11.9).

11.2. : Show that $\overline{\psi_k^+} = \psi_{-k}^-$.

11.3. : Let V and Δ be as in Proposition 11.3. Show that

$$(x,k) \mapsto \psi_k(x) - \psi_k^0(x)$$

is uniformly continuous on $\mathbb{R}^n \times (\Delta \times S^{n-1})$.

12.1. : Let H be the self-adjoint operator defined by

$$(H\tilde{f})(k) = \psi(k)\ \tilde{f}(k)\ .$$

where $\psi : \mathbb{R}^n \to \mathbb{R}$. Prove that (12.3) holds for each $f \in \mathcal{H}_c(H)$.

12.2. : If $\phi \in L^1(\mathbb{R})$ and $A = A^*$, one has

$$\tilde{\phi}(A) = (2\pi)^{-\frac{1}{2}} \int_{-\infty}^{+\infty} \phi(t)\ e^{-iAt}\ dt\ .$$

13.1. : Prove Lemma 13.2.

13.2. : Prove (13.7).

13.3. : If $D(H) = D(K_0) \subseteq D(V)$, then

$$\|(R^o_{\lambda-i\varepsilon} - R^o_{\lambda+i\varepsilon})\ V\ R_{\lambda+i\delta}\| \leq M(\delta,\varepsilon) \quad , \quad \forall\lambda \in \mathbb{R}\ .$$

13.4. : Show that the unitarity of S implies that of $S(\lambda)$ for almost all λ.

13.5. : Show that the operator $P(\phi)$ defined in (13.22) is an orthogonal projection and its range is the subspace $N(\phi)$ introduced in (13.23).

13.6. : Prove that $f(\lambda\ ;\ \omega' \to \omega) = f(\lambda\ ;\ -\omega \to -\omega')$.

14.1. : Let ϕ be as in Proposition 8.2, and let $M_\phi(\lambda)$ be the closure of $\hat{M}_\phi(\lambda)$ as in Proposition 8.2. Prove the following :

(a) if $g \in D(K_0)$ and $f \in D(\phi(Q))$, then

$$(g,\phi(Q)f) = \int ((\mathcal{U}_0 g)_\lambda\ ,\ M_\phi(\lambda)f)\ d\lambda,$$

(b) $[\mathcal{U}_0\ \phi(Q)f]_\lambda = M_\phi(\lambda)f$ for almost all $\lambda \in \mathbb{R}$.

Bibliography

BOOKS

[A] AGMON, S. : Lectures on elliptic boundary value problems. Von Nostrand Studies # 2, (1965).

[AG] AKHIEZER, N.I. and GLAZMAN, I.M. : Theory of linear operators in Hilbert space. (English translation), vols I and II, Frederick Ungar, New York, (1963).

[AS] AMREIN, W.O. and JAUCH, J.M. and SINHA K.B. : Scattering theory in quantum mechanics. W.A. Benjamin, Inc. (1977).

[DS] DUNFORD, N. and SCHWARTZ, J.T. : Linear operators. Parts I and II, Interscience, New York (1958 and 1963).

[H] HÖRMANDER, L. : Linear partial differential operators. Springer, New York, (1969).

[J] JAUCH, J.M. : Foundations of quantum mechanics. Addison-Wesley, Reading, Massachussets, (1968).

[K] KATO, T. : Pertubation theory for linear operators. 2nd edition, Springer, New York, (1976).

[LL] LANDAU, L.D. and LIFSHITZ, E.M. : Quantum mechanics, non-relativistic theory (English translation), Pergamon, Oxford (1958).

[LM] LIONS, J.L. and MAGENES : Problêmes aux limites non homogênes et applications, vol. 1, Ed. Dunod. Paris (1968).

[N] NEWTON, R.G. : Scattering theory of waves and particles, Mc Graw-Hill, New-York, (1966).

[R] ROYDEN, H.L. : Real analysis, Macmillan, New York, (1963).

[RN] RIESZ, F. and SZ. NAGY, B. : Functional analysis (English translation), Frederick Ungar, New York, (1965).

[RS] REED, M. and SIMON, B. : Methods of modern mathematical physics, Vols I, II and III, Academic Press, New York, (1972, 1975, and 1979).

[S] SCHWARTZ. L. : Cours d'Analyse de l'Ecole Polytechnique. Vols I et II, Hermann, (1967).

[SI] SIMON, B. : Quantum mechanics for Hamiltonians defined as quadratic forms. Princeton Univ. Press, Princeton (1971).

[T] TITCHMARSH, E.C. : Eigenfunction expansion associated with second order differential equations. Parts I and II, Oxford Univ. Press, Oxford (1962 and 1958).

[W] WATSON, G.N. : Theory of Bessel functions. Cambridge Univ. Press, Cambridge, (1944).

[WW] WHITTAKER, E.T. and WATSON, G.N. : A course of modern analysis. Cambridge Univ. Press, Cambridge, (1963).

ARTICLES

[1] AGMON, S. : Ann. Scuola Normale Sup. Pisa, Series IV 2, 151 (1975).

[2] AGMON, S. : Séminaire Goulaouic-Schwartz, Ecole Polytechnique, n°2, 1978-1979.

[3] ALSHOLM, P. : J. Math. Anal. 59, 550 (1977).

[4] ALSHOLM, P. and KATO, T. : Proc. Symp. Pure. Math. vol. 23, Amer, Math. Soc., 393, (1973).

[5] ALSHOLM, P. and SCHMIDT, G. : Arch. Rational. Mech. Anal. 40, 281, (1971).

[6] AMREIN, W.O. and GEORGESCU, V. : Helv. Phys. Acta 46, 635 (1973).

[7] AMREIN, W.O. and PEARSON, D.B. : J. of Physics A, vol. 12, 1469, (1979).

[8] AMREIN, W.O. and PEARSON, D.B. : Ann. Inst. Henri Poincaré, A, 30, 89 (1979).

[9] AMREIN, W.O. PEARSON, D.B. and WOLLENBERG, M. : Evanescence of states and Asymptotic Completeness, Helv. Phys. Acta, HPA, 53, 335, (1980).

[10] BERTHIER, A.M. : On the point spectrum of Schrödinger operators, Ann. Scient. Ec. Norm. Sup., 4ème série, t. 15, 1982.

[11] BERTHIER, A.M. : C.R.A.S. Paris, t. 290, Series A, 393 (1980).

[12] BERTHIER, A.M. : On the existence of the modified wave operators for long range potentials. Preprint (1978).

[13] BIRMAN, M.S. : Math. Sb 55, 124, (1961) (Amer. Math. Soc. Trans. 53, 23, (1966).

[14] BUSLAEV, V.S. and MATVEEV, V.B. : Theor. Math. Phys. (English translation) 1, 367, (1970).

[15] CWICKEL M. : Ann. Math. 106, 93, (1977).

[16] DAVIES, E.B. : On Enss' approach to scattering theory, Duke Math. J. 47, 171, (1980).

[17] DEIFT, P. and TRUBOVITZ : Inverse Scattering on the line CPAM to appear.

[18] DOLLARD, J.D. : J. Math. Phys. 5, 729, (1964).

[19] ENSS, V. : Comm. Math. Phys., 61, 285, (1978).

[20] ENSS, V. : Ann. Phys. New York, 119, 117 (1979).

[21] FARIS, W.G. : Self-adjoint operators, Springer Lectures Notes, vol. 433, (1975).

[22] GINIBRE, J. : La méthode "dépendante du temps" dans le problème de la complétude asymptotique, RCP 25, Strasbourg, (1979).

[23] HÖRMANDER, L. : Math. z. 146, 69, (1976).

[24] JANSEN, K.H. and KALF, H. : Comm. Pure Appl. Math. 28, 747, (1975).

[25] KATO, T. : Comm. Pure Appl. Math. 12, 403, (1959).

[26] KATO, T. : Math. Ann. 162, 258, (1966).

[27] KURODA, S.T. : An introduction to scattering theory, Lectures Notes, Aarhus Universitet, n°51, (1978).

[28] LIPPMANN, S.A. and SCHWINGER, J. : Phys. Rev. 79, 469, (1950).

[29] MOURRE, E. : Comm. Math. Phys. 68, 91, (1979).

[30] NEWTON, R.G. : J. Math. Phys. (1980).

[31] PEARSON, D.B. : Comm. Math. Phys. 40, 125, (1975).

[32] PERRY, P.A. : Mellin transforms and scattering theory, Duke Math. J. 47, 187 (1980).

[33] RUELLE, D.: Nuovo Cimento, Vol. LXI A, n°4, 655, (1969).

[34] SCHWINGER, J. : Proc. Nat. Acad. Sci. USA 47, 122, (1961).

[35] SIMON, B. : Duke Math. J. 46, 119, (1979).

[36] THOMAS, L.E. : Comm. Math. Phys. 33, 335, (1973).

[37] WIENER : Duke Math. J.5. 1, (1939).

[38] YAFAEV, D. : On the proof of Enss of asymptotic completeness in potential scattering, preprint, Institut Steklov, Leningrad, (1979).

Subject index

Notation index